Walk Among Them

by

Rick Friday

edited by

Anita Gaunt

First paperback edition April 2021

Book design by Rick Friday

ISBN 978-0-578-99074-3 (paperback)

www.fridaywritings.com

Dedicated to my darling wife
and to the spouses of GIs
everywhere

Foreword

I was writing for a small national magazine and my editor asked me to do a piece on a chaplain who had been deployed to a combat zone. I contacted the Public Affairs office of the 39th Infantry Brigade of the Arkansas National Guard headquartered at Camp Robinson, Arkansas. After researching my credentials, they put me in contact with Chaplain (Maj) Jeremy Miller.

I interviewed him and wrote the article. However, during the interview, I discovered there was a lot more to this story than an 1100-word article. His family, his journey, and his experiences simply required a much deeper telling. I could not shake the thought there was a much broader story to be told.

About a month after the article was published, I approached Jeremy with the idea of writing a book diving into his life and his experiences. I asked him to discuss this with his wife and family and to pray about it.

Shortly after discussing it with him, Jeremy called me and said, "Let's do this." So began a journey that involved numerous interviews, spending several days with him just following him around, and developing a very good friendship with this godly man.

This book isn't about a man charging enemy forces with only a Bible in his hands. Harrowing stories of war can be found in any library or bookstore. This book is about the journey of a man following God. It tells the story of a change that God made in this man and the charges He placed in the man's heart. This is a story of that man's calling by God that continues to impact lives today.

Introduction

As he stepped down from the aircraft after conducting his first ramp ceremony (the ceremony to send a deceased soldier on their final flight to the United States from a combat zone), this prayer went through Chaplain Jeremy Miller's mind: "God, I pray that I just honored Sgt. McPherson the way he would want to be honored. And if something happens to me and I don't make it home, I hope somebody honors me in that way." The fear of not having done enough for this soldier was on the forefront of Chaplain Miller's mind.

As he looked out to the distant edges of the base, he had another thought: "This could have been me. I'm not kicking in doors or dealing with direct fire but there is the chance this could happen to me." The true realization that he was now in a combat zone hit him squarely in the face. He became completely aware he was one rocket, one mortar round, or one bullet away from meeting God face to face. He instantly thought of his wife and how she would react when the death notification team showed up in front of his house. Having done death notifications before and knowing how they work, he knew this could be happening right now to the family of Sgt. McPherson. He knew there was someone, including a chaplain like himself, driving to that house, knocking on that front door, and delivering the worst news a wife, a mother, and a family can get, with the distinct probability that someone was holding a baby or young child as this news was delivered.

All of these things were running through his head. He was discovering what everybody who has been deployed into a combat zone discovers; there is this constant rush of emotions that you do not have time to sort out. They are hitting you all the time. These emotions, fear, dread, and relief ran though his mind. He looked into the mountains surrounding the base and the thought occurred, "My enemy is probably watching us right now."

As he turned away from the aircraft, the soldiers on the ramp began approaching him and saying, "Great job, Chaplain. We really appreciate you." He walked among the soldiers lingering on the tarmac as the plane readied to leave, and they continued to come by and speak to him, causing this rush of emotions to begin to subside. He was comforted by the fact these soldiers are comforted and appreciated what he just did and said. He silently acknowledged to God, "I now truly know why you have sent me here, and I will do Your work here in Afghanistan."

Chapter 1

The Beginning

Jeremy Miller was born in 1981 in Conway, Arkansas. It would be nice to say he was born to a "normal" family, but that wasn't the case for him or even his parents, Johnny and Sue. Don't take that the wrong way. They were normal in the fact there was a dad, a mom, and siblings. It was how the Miller's made their living that was out of the ordinary. There have been a lot of little boys who have thought, dreamed, or talked about running away to join the carnival, but that wasn't the case for Jeremy. He was born into it. You also can't confuse the circus with a carnival, either. There were no elephants, trapeze artists, or even a ring master. There were no big top tents packed with peanut eating crowds awaiting the horses and acrobats. No, his family, dating back to his father's grandparents, had owned, and continue to own, a traveling carnival. Diesel generators rumbled over the sounds of carousel music and crowds lined up for everything from a corn dog to a swing ride. At the time his grandfather owned it, the carnival was named Miller Gala Exposition. When his dad and uncle bought the carnival from their dad, they changed the name to Miller Brothers Midway. Today, the carnival is directed by his brother who is taking Miller's Spectacular Shows to new heights. The families focus can simply be summed up in the words "they sell fun."

Jeremy's mother's parents worked on the carnival, as well, and this is where Johnny and Sue met. This was the only way of life Jeremy knew until he graduated from high school. Jeremy's dad and all of his dad's siblings started managing carnival rides at a young age. His dad told Jeremy he could remember the days when mules provided the power for the rides. Since both sides of the family worked the same carnival, it was inevitable that Johnny and Sue would meet, fall in love, and get married.

The entire family worked and traveled together across the mid-south, an area encompassing Missouri, Oklahoma, Arkansas, Texas, and Mississippi. Jeremy has photos of his grandfather holding the microphone for Jerry Clowers, a famous entertainer from Mississippi. As Jeremy said, "Nobody retired, nobody quit working. They just kind of died in place." His grandparents on both sides didn't really retire. When they quit traveling with the carnival, they continued in a "working retirement." They became trucking "flaggers," better known as escorts, for the oversize loads.

Marvin, who was Sue's dad, credits being able to drive a truck with possibly saving his life. He enlisted in the Army during WWII and was an infantryman. He was stationed in England and was preparing for the D-Day invasion. Somebody came through the ranks one day looking for truck drivers due to a shortage. When his company was approached, he volunteered stating "This is all I've ever done". In fact, he carried an over the road (OTR) large truck license. He spent his time driving trucks back and forth to the docks and ships with equipment and material. This led to him driving a deuce-and-a-half around England as the invasion took place.

Burt, who was Johnny's dad, was the same type of hard charging individual and was a really old school type of person. When they arrived in a town, he would walk the lot detailing where every ride and vendor was to set up. He knew precisely how large each ride was and where it would need to go and he knew how many feet each of his steps measured. He would walk the lot (area of the carnival) with a six pack of beer. When he reached the end of where a particular ride would need to go, he would finish a beer, crush the can and throw it on the ground as a marker. "Edge of the trailer right here," he would say, and that's where the trailer would back up and off-load the ride. As soon as the ride was in its place, crews of men would work to arrange the steel pieces and parts. Others would run the power lines, and within a day, the location was alive with rides and games seeking to put a smile on every face. Burt was the one who built the carnival into a thriving business.

Johnny and Sue have been married over fifty years and have only recently retired, letting Jeremy's brother and his kids manage the business. The range and territory expanded under Johnny's leadership, and they now travel as far west as New Mexico and as far north as Michigan. This work ethic has been instilled in all of their children and is being passed on to each of their grandchildren. To this day, there are only about four adult family members in the entire family who do not work in the carnival business. One is a police officer, two are in the military (including Jeremy), and Jeremy's sister is an English teacher.

Jeremy was born in the carnival off-season. Sue was pregnant with him the season prior while working Sue's Corndog Stand. Every day over the course of that summer, she spent hours standing in front of a corn dog fryer serving the crowds. The season ended and they returned to Greenbrier, and Jeremy was born prior to Thanksgiving. Four months later, the carnival hit the road, and so did Sue and Jeremy. He was the baby of the family by a long shot. His brother Freddy is fifteen years older and his sister Susan is thirteen years older. As Jeremy jokes, "I am a November baby so I probably have Valentine's Day to thank for that." His brother married right out of high school so when his sister got old enough, he stayed with her when the family traveled, continuing to live in Greenbrier. His sister was going to college at the University of Central Arkansas, majoring in education to teach English. Susan was the first Miller to go college, although many more would follow.

Sue was very much a Christian woman, dedicated to her faith and attending church, as well as raising her children in church. Johnny, at that time was not a Christian and did not go to church. During the off-season, on Sunday, Johnny would head to the shop and work on equipment. One of the things he taught Jeremy was, "If you are going to tithe to the church, you have to make the money first." As Jeremy got older, it became his option on whether to stay and work with dad or go to church with mom. If he worked with Johnny, he got to sleep in, and as he got older, that was a priority.

Jeremy was exposed to church and Christianity at a young age. At six years old, he remembers going to church and the Sunday school teacher coming in with a flannelgraph. A flannelgraph is a storytelling system that uses a board covered with flannel fabric, usually resting on an easel, with its primary use being a storytelling medium. One could attach pictures to the flannel to help illustrate the story. This particular one had a picture of a man on the cleft of a rock. Below the picture she put fire and a picture of people burning in the fire. She asked the question, "You don't want to go to hell and burn there for all eternity?" Of course, all the six-year-olds in the class room said no. So she said pray a particular prayer and they all did. After that, she walked them out to the church body and presented them as new Christians. After doing this, they began to talk about baptism and one of his great friends told him he was fairly certain they drowned kids in the baptistery. This scared him and he told his mom he didn't want to be baptized, though he still went to church as expected and continued to hear the teachings. For a long time, this “prayer” of giving his heart to Jesus in order to avoid hell was what he based his Christian identity on.

Chapter 2

The Carnival

Jeremy worked every day in the carnival during summers as a young child. He would work with the game and food vendors and mainly help out wherever he was needed. The neatest job was going around and picking up money from ticket boxes. No vendor liked to keep a large amount of money in their booths, so as Jeremy would come around, they would give him a certain amount of money. For example, the ticket sellers would give him half of the twenties and all of the hundreds. He put this money into a bank-type bag and then put the bag into the front of his pants. He would cover it with his shirttail and take off through the midway to the office. Now, it sounds dangerous having a kid walking through the midway with this amount of cash, but no one was aware he had these large amounts and he was totally unnoticed by the patrons. His parents would tell him to look down while he walked to make sure he didn't drop anything, and he would go straight to the office. He would give the money to his mom, who also functioned as the accountant for the carnival.

Other jobs he had were to run tickets to the booths, serve drinks in the food wagons, and get ice. Basically, he was a gofer. This continued until he was about fourteen. At that time, his responsibilities grew. The carnival had expanded to the point where they had to be in multiple locations at the same time. His brother was already managing his own unit and Jeremy would travel with his brother or with his parents, depending upon who needed him and his newly acquired crew the most. This experience taught him to manage people, resources, and time at a young age.

His crew's responsibility was to move a couple of rides: the Super Slide and the Granny Bugs, as well as a couple of other small pieces. After moving and setting them up, they were responsible for manning and operating the rides. The orders from his dad and brother were to make sure the crew was fed, paid, clean, and in the right uniform. Jeremy was a crew chief. Quite a responsibility for a young man of fourteen, but it was what every family member had done at that age.

With the exception of one week during the summer, Jeremy was not exposed to any church activities. A small Baptist Church in Eldon, Missouri, would hold a Vacation Bible School for all of the kids in the carnival, and the church would come out to the carnival to pick them up. Jeremy has great memories of this one exposure to church during carnival time. Because he was always on the road, he never got to go to church camps or mission trips or enjoy any of the other summer church youth activities.

When the family came off of the road (off-season), Johnny would decorate the home with Christmas lights. Now, don't misunderstand this statement. We are not talking about just putting up some Christmas lights. Johnny did things in a big way. People would drive from considerable distances to see this light display, and they would pile up in the front of the house in rows of traffic. It was massive. There were two-story tall Christmas trees, a tunnel of Christmas lights to drive through, and untold numbers of displays. The entire lighting system was run on the same generators that powered the carnival. The family lived to bring joy to others, and this was the way they did it when the carnival was not running.

Of course, being a young boy, there was some always some kind of mischief to be gotten into. Jeremy and his buddies would do a thing called "purse fishing". This consisted of a fishing pole, fishing line, and a woman's purse. Doesn't take much to see where this is going. They would tie a purse to the line from the pole and lay the purse by the side of the road. They would have a little Monopoly money sticking out of the purse to further entice somebody to get out of their car and go for the purse. The boys would run back and hide inside one of the two-story Christmas trees. As people would stop to admire the lights, they would notice the purse on the ground and get out of their car. As soon as they got near the purse, the boys would reel it in, causing a number of people to chase the purse. They were made to quit when they nearly gave one woman a heart attack from running so hard chasing the purse.

The family would also put on a Christmas gift exchange for children. They would travel to Hot Springs, Arkansas, and would host between seventy-five and one hundred under-privileged children. They would bring presents and fruit and would pile them high around a tree. One of Johnny's workers would dress up as Santa Claus, Jeremy would dress up as an elf, and they would bring all of these children to the tree over the course of one or two days. Johnny continues to do both the lighting display and the children's gifts.

Besides getting his own crew to manage at fourteen, Jeremy also received a hardship driver's license at that age. The excuse was he had to get to and from school because his parents were out of town and his sister was working. This was also the time when Jeremy began living by himself. When Jeremy left the carnival to come home for school, his parents sent him home with money. His sister would check on him occasionally, but primarily he lived by himself. He had a white Chevy pickup, which happened to be his dad's old work truck. The truck was made with steel instead of aluminum, and as his dad handed him the keys, he said, "If you get in a wreck with this truck, God help the other person."

Jeremy had three friends growing up: Jonathan, Will, and Jake. Those were his running buddies throughout his school life. Having a truck and driving at this young age made him extremely popular among his peers. They didn't have licenses to drive, let alone a vehicle. However, he didn't always make the best decisions with his truck.

For example, they would do Future Farmers of America (FFA) fruit sales, and after the sales, Jeremy and his buddies would deliver the fruit to the buyers. Old fruit was thrown into the bottom of the bed of the truck. One of his buddies came up with the bright idea of driving by and throwing eggs at one of his teachers' houses. Jeremy thought that was a great idea (it wasn't), and he drove them to this teacher's house where he and his buddies promptly began throwing eggs. As they were driving away, they noticed they still had all of this bad fruit in the bottom of the truck. They looped around the street and went back to the house, which was their second really bad idea of the day because everybody knows you don't make a second trip. As they drove up and started throwing fruit, they noticed the teacher on the front porch with the phone in her hand. Turns out, she had the police on the line and was able to give them a great description of the truck.

Greenbrier is a small town, and teachers knew which student drove which vehicle. By the time Jeremy got back to his house, there was a police cruiser sitting in his driveway. They came up with another brilliant idea, their third of the day, to try to outrun the police to the county line. They drove to one of Jeremy's friends' house and lo and behold, there was another police cruiser sitting there. Realizing all was lost, they pulled over. Jeremy still remembers the officer saying over the loud speaker: "Jeremy, Jake, and Jonathan, could you please get out of the truck?" They were amazed the police knew who they were. Jonathan tried to hide under the tarp. Not being fooled at all, the officer stood by the bed of the truck and said, "Are you going to come out from under the tarp?" Jonathan came out from under the tarp with his wallet and asks the officer, "How much to make this go away?" Jeremy nearly fell over, knowing they were going to jail for sure. They did get a ride to the police station and his truck was impounded. He had to call his sister to come pick him up and also go with him to get his truck from the impound lot.

There is also another advantage to living in a small town. The teacher didn't file charges, and they were allowed to go free. No one was charged, but the boys did have to publicly apologize to the teacher. They got a slap on the hands with the promise they wouldn't do anything like that again. They lived up to this promise, and this was the most serious trouble the group got into during their junior high and high school years.

All four of the guys left Greenbrier after high school, but Jeremy, Will, and Jake have all gravitated back to the area. Jeremy still sees Will and Jake occasionally. The most recent time together was at the Daddy/Daughter dance at the Greenbrier school where all three of them showed up, each with their two daughters. Jonathan, however, has taken a different tack. He moved to Colorado and has settled into a life he enjoys living in that area.

Chapter 3
Football

As stated earlier, shortly after his birth, Jeremy's parents took him on the road with the carnival and he traveled with them until he reached school age. Once he reached school age, his mother would come back early enough to get him registered and started in school. He would stay with his grandparents or friends unless his mother was needed for some important event. This happened until the carnival left the road in October or November. Once the school year was over, Jeremy would again hit the road with the carnival. Most childhood sports at that time were either started or were played while Jeremy was on the road. This is one reason why he never developed an affinity for other sports. He would leave the last day of school and come back the day prior to school starting or, as he got older, the day before football practice would begin. By the time he reached Junior high school age, this was a well-established pattern. Football became his first true love.

Football was Jeremy's identity because it was a team sport and he could hit people, as well as protect people. From that, it built into being his team and an "us against the world" mentality. Jeremy would be like a sponge around his coaches. Coach Edwards and Coach Logan were his junior high coaches and built into Jeremy a "seize the day" mentality. These two men invested more time in Jeremy's life than his own dad did, due to the circumstances of his dad being on the road with the carnival. In fact, these were the two men who came to him in school the day one of Jeremy's grandfathers died. They broke the news to him and helped him get through that day. They were able to instill in the players the will to want to give one hundred percent all the time, and they had a pretty good team in junior high. Jeremy established his offensive position during these years. He was the center. He handled the ball on every play and played this positon until he graduated from high school. Defensively, he played defensive tackle or defensive end. There were only twenty five players on the team so everybody played both ways.

Between his ninth grade and sophomore years, Jeremy began to put on size and weight. Again, don't misunderstand this. 6'1" and 185 pounds was as big as Jeremy ever got in high school, but he was quick and even though he would often go against players much bigger, he was able to use his quickness to his advantage. It was also the time when the coaches wanted him to play more on the defensive side of the ball.

In junior high, no one knew what to expect. Coming from youth football, all of the players had played on different teams. Seventh grade was the point where they all came together. The first rude awakening came when they all had to have a physical. The physicals were given in the old field house. The doctor came in and checked each player, not individually but as a group. Without going into detail, it was a completely new and different experience for these boys.

They moved from there to the weight room and the coaches began to teach the boys how to work out. They would crank up the radio to one of the local classic rock stations and get down to business. Jeremy distinctly remembers their first power lift. It was a forty-five pound bar with no weight on it. There were guys struggling just to get the bar off the ground, much less lift it above their heads. Jeremy, having grown up doing heavy lifting in the carnival, had zero trouble with the bar. He began to feel at home with the weight and workout room. After lifting and working out, at the end of the day, the coaches would make them run to the stop sign a half mile away and run back. It was kind of dangerous because the stop sign was on a major highway. The guys would run to the stop sign, hit it, and run back. Luckily, none of them got run over.

Jeremy's dad had traded in the work truck for a single cab Chevy pickup and he started his sophomore year with the new truck. For a new sophomore, he was the man and tricked out his new truck. He lowered it, put in a new bed liner, installed a ground effects light kit, and named it Miller Time, which also had become his nickname.

Jeremy had been a starter in football since his seventh grade year. The high school season hadn't started, and he didn't have a starting spot on the high school team. As is the case in all high schools, the sophomores were treated unmercifully by the upper classmen. They had to take out the trash, carry all of the equipment for practice, and basically were treated like second class citizens. And this was to be expected. As the coaches said, they had to earn their place around the team. That didn't mean Jeremy just lay down and let people walk over him. He practiced and played like he did in junior high. He scrapped and played hard. He went after everything like it was his. He was put on the scout team defense and went up against the starters. He went against the same guys every practice. He said they were huge, not just in size but in everything. They were seniors and had been through the wars. They were bigger than he was, but he still thought he had to get to these guys.

Through the first four games of the season, because of his practice play, he was getting reps during the game and those reps were increasing with each game. One day in practice, the coaches had Jeremy and the starting defensive tackle go one on one. It's a little drill called Bull in the Ring, and you go at each other until one quits. Jeremy saw this as his chance to get the starting spot, and in his mind, he knew he could beat the other guy. After this drill, Jeremy became the starter and never relinquished his spot.

This didn't sit well with the seniors on the team. The next practice after Jeremy was announced as the starter, there was a line drill, defense against the offense. As the drill started, a couple of the seniors pinned Jeremy to the ground and they stomped on his hands with the thought process of breaking his hands so he couldn't play and the senior could get his job back. Jeremy still has those scars on his hands from that day. After practice, with his hands bleeding and bandaged from being stomped on, he went to the store and bought padded gloves. Next practice, he wrapped his hands up like a boxer and put on these padded gloves. Now stomping on his hands will do no good, but it allowed Jeremy to come in with everything he had, throwing punches and grabbing face masks. He was not going to be intimidated. It made him mean, and he carried that meanness over to the games. It only enhanced his position as a starter.

This went on for a couple of weeks. One day, as he was taking out his part of the trash, the senior quarterback was waiting for him by his truck. Dread came over Jeremy because all he could see was trouble. The quarterback approached him and as Jeremy opened the truck, the quarterback told him to move over because he had never sat in his truck. They started the truck and Jeremy had "Magic Carpet Ride" playing in the CD player. (This was the music Jeremy grew up with traveling on the carnival with older people.) He was concerned this might be something else for them to pick on because it was the era of the boy bands. However, the quarterback was impressed by the music, and this began the process of Jeremy being accepted by all of his teammates. There were no more issues with guys wanting to get him out of the lineup. His play improved, and he finished the year as the starting defensive tackle. Football became his God.

Chapter 4
FFA

As Jeremy continued to grow, he found the second love of his life: FFA. The FFA teacher was Mr. Patrick Breeding. Jeremy started like all freshmen do, at the bottom. He took the basic agriculture class but it was also a chance to get into the shop and get to work with tools. Jeremy isn't a gear head, but growing up on the carnival, he had to know his way around tools and machinery. When his sophomore year started, Mr. Breeding put him on the land team and the mechanics team.

FFA offered a large number of opportunities. Jeremy wasn't a great student, but he was a good student. He didn't love school and he wasn't focused on getting great grades, but he did all right. School gave him the opportunity to play football and be in FFA. One of the greatest successes in FFA was learning how to land judge. The Greenbrier team placed in state competitions every year, were state champions his junior year, and were able to go to nationals in Oklahoma City.

Land judging consists of testing the soil for its possibilities. A pit would be dug and each team would begin an examination of the soil in the pit. Outside of the pit, the slope and drainage of the land was examined. Inside the pit, the quality of the soil was examined for a number items, basically checking to see if the soil was fit for agriculture. The composition of the soil was analyzed and the findings were presented to the judges. Winning state and going to nationals was a big deal. At nationals, all of the team members competed and the lowest score from all of the team members was dropped.

There was an individual on the team named Joey. Joey didn't have a lot of options and opportunities in life. In fact, going to Oklahoma City was Joey's first trip outside of the state of Arkansas. He would ask people from Oklahoma if he could take a picture with them because he had never met anybody from Oklahoma. Joey worked hard all year, but usually Joey had the lowest score on the team and his scores were dropped. This didn't happen at nationals. After climbing into the pit and doing their analysis, all of the team members' scores were compiled. Joey's score placed and Jeremy's was the score that was dropped. Needless to say, Jeremy was kind of shocked, but this became the highlight of Joey's life. He was on cloud nine. It wasn't about beating Jeremy but about the fact that he got on a big stage, in his senior year, and he shined. Jeremy wasn't sad because he got to see the joy it brought to Joey and he got to celebrate with his teammate. Sadly, two years after graduation, late one night, Joey fell asleep at the wheel of his car, ran off of the road, and hit a tree, killing him instantly.

Since he had gone to nationals, Jeremy could no longer be part of the land judging team. So, he became an officer in the FFA and became part of the parliamentary speaking team, and this was where he learned the principles of articulation and the ability to do public speaking. He wasn't able to enjoy the success he had with the land judging team, but he discovered a gift he didn't realize he had. He became very comfortable with parliamentary speaking, and it has proven to be very beneficial in his life. Being on this team also provided him with the opportunity to travel. State competition was in Fayetteville, Arkansas, and he remembers those trips as being a lot of fun. FFA was good to Jeremy and really helped shape a lot of who he is today.

Chapter 5

Sixteen Years Old

Football success wasn't the only thing Jeremy would find the year he turned sixteen. Jeremy's junior year in high school was where he discovered he had leadership qualities. His football ability began to shine, and he was beginning to be noticed around the school. In addition, like any other sixteen-year-old young man, he began to notice girls a lot more and they began to notice him. This became readily apparent the summer prior while he was working on the carnival. When the carnival came to a town, he looked for two things; where the nearest football field was and where the girls hung out in town. He would work hard at the carnival, but his free time was spent chasing skirts and working out on the football field.

Starting his junior year, he began to get a "me against them" attitude. This came about due to the fact he was basically raising himself, living alone, and being totally responsible for his life. In his mind, since he could do all of these things, who had the right to tell him what to do? This led to his not having a lot of respect for his principal and some of his teachers. However, his English teacher, Miss Sherry Williams, made a deep impression on him. Miss Williams began to teach her class about civil disobedience. There was a heavy dosage of Martin Luther King and how he had changed the world through non-violent action.

Someone in the high school administration decided that salt shakers needed to be removed from the lunchroom. This was something hard to understand for the students. High school cafeteria food is not always the best tasting, and it is helpful to put a lot of salt on it. By not having salt, the food was just not very good to eat. Jeremy's attitude about being told what he can and cannot do kicked in along with his new found leadership qualities. Jeremy organized what is now referred to as the Salt Strike. Miss Williams instructed him in this way: As long as you don't break any rules, you can protest this matter. So they did, and she began to help as much as she could and still keep her job. Jeremy was smart enough to know holding up signs would have no effect. The way to have effect was to hit the school where it hurt: in its pocketbook. Instead of purchasing lunches every day, the students began to bring their own lunches and packed their own salt. The lunchroom now was preparing a bunch of food that was not being sold. After three days, the principal called them into the office and told them to just stop. The students countered with, "You can't tell us not to bring our lunches."

The school office was receiving calls from parents curious about why their children were wanting to have lunches prepared at home simply because they couldn't get salt at school. Students responded to Jeremy and it became school wide. Miss Williams told him they were having teachers meetings about this single subject, and so it continued for the next two weeks. Jeremy discovered students were listening to him and he had an audience. He also began to faintly hear God's voice about leading, but didn't pay it much heed at that time. He also understood he needed to make sure he didn't mess up this newfound skill and talent he had been given. He always made sure he was leading in the right way, or trying very hard to do so. The students won, and they were given their salt back. The principal didn't much care for Jeremy the next two years, but it became water under the bridge between them later in life.

He had also gotten pretty good at what he calls "faking the funk" concerning his church. He would do things on Saturday night, hiding them from his mom, then get up on Sunday morning and go to church like nothing was wrong. He actually thought that was what a Christian did because he wasn't alone in these actions. Students he would party with on Friday and Saturday night were in his youth group at church.

Going back to the girls, once school began, Jeremy began dating. There were a couple of girls in school he would take out occasionally but there wasn't anybody he dated exclusively. At the winter break, he was informed about a new girl in town. She was working at a local department store and Greenbrier being a small town and all, Jeremy just had to see who she was. He found out her dad had moved here and their family was basically starting over after selling their farm. Her dad went to work at UPS, working the night shift in order to attend the University of Central Arkansas during the day. Everybody in the family had to work to make ends meet.

Off to the store he went. It was not hard to spot the new girl in town. Not only was she new to town, she was, in Jeremy's terms, strikingly beautiful. He jokes that the first time he met her, she was checking him out, literally. She was the cashier in the store. When he got to her register, he was dumbfounded. Going through the checkout line, he could barely get any words out of his mouth. Picture this guy who has been organizing protests and people and suddenly he's struggling to even put two words together. He left the store and took the groceries home to his mom who informed him he had forgotten the milk. So, back to the store he went.

He ended up making four or five trips to the store that day simply to go through this girl's lane. Each time, he got up a little more nerve, and by the last time, they were having a real conversation. He found out her name was Beth Bennett and they were in the same grade. She had broken up with a guy from her previous town and was starting fresh. In addition, she was a true born-again Christian. Beth also did her research on Jeremy. She found out he was a jock and liked to party, and that immediately turned her away from him. It got worse because her family began attending Jeremy's church, Faith Baptist Church, and he saw her constantly. As the spring semester started, he discovered she was in his computer class. They flirted like crazy but she would not go out with him. She started dating a senior, and Jeremy just couldn't understand why. His crush intensified, but he got nowhere with her.

Jeremy turned to his youth minister for guidance, but got guidance in a way he didn't expect. All his youth minister knew was, according to the church roles, Jeremy was a saved member of the church and was counseled as such. His youth minister knew Jeremy could speak publicly, so he started mentoring him to become a preacher and actually allowed Jeremy to preach on a Sunday night to the church. So picture this: Here's a guy who is partying on Friday and Saturday nights, who has a bad attitude as far as authority is concerned, and is definitely not a saved individual preaching to his church with his youth group in attendance, a lot of whom were just out the previous night partying with him. He ended his junior year a very confused young man.

Chapter 6

Senior Year

Jeremy's junior year of high school football did not turn out as he expected or desired. His team ended the season with a losing record, they lost to their rival, and all of this left a bad taste in his mouth. On a personal note, Jeremy experienced recognition for his play and effort. As a junior, he was named to the All-Conference team, one of only two players from his team to be recognized in this manner. His junior year also showed him if he wanted to achieve his dreams of playing on the next level, he needed to step it up.

When the school year ended, he picked up the routine he had followed all of his life, going back on the road with the carnival. However, this summer would be different. He was given more responsibility in every aspect of his work. He was traveling back and forth between his parents' unit and his brother's unit. He was given more people to supervise and three to four rides to manage. While growing in size and maturity, Jeremy's primary growth this particular summer was in aggression. This growth came about due to his adherence to achieving his goal of becoming a better football player.

Regardless of where the carnival was located, he would find a workout facility and would make it a very important part of his daily routine. Every morning, he would go to these facilities and, in Jeremy's own words, "work out like crazy." He would spend two or three hours working out to start the day before heading back to the carnival around noon. He would prep and do maintenance on his rides, and the carnival would open at four in the evening. The carnival would close somewhere around ten in the evening, and cleanup would begin, meaning it would be after midnight before all was said and done for the day. This makes for long days.

With this schedule, the groundwork was laid for his aggression growth. Added to this groundwork was the hot summer days of the South. Unless you have lived it, it is hard to put into perspective. Summer days in Mississippi, Louisiana, Arkansas, and Texas can be some of the most miserable anywhere. It's not the heat alone that makes it so unbearable. When you add 50% plus humidity to 95 degree plus heat, it can become unbearable. Most of the guys who were working for Jeremy were of legal drinking age and after working a long, hot day in the sun, their first inclination after getting off of work was to get an ice cold beer, and they never stopped at one.

They also weren't individuals of the highest character. When getting into an area, Jeremy and his brother would go on a "mission run." What this entailed was taking a fifteen-passenger van and heading for the nearest mission work facility. They would ask the individuals staying at the mission if they wanted a job for a short time. The carnival would provide salary, room and board, and uniforms to wear, and they wouldn't quit hitting these places until they had filled up the van.

Add together tiredness from working in the hot sun and alcohol, and the usual result was a fight of some sort around the carnival complex, usually waking those who had not been out drinking. Jeremy's brother's method of handling these instances was, "Just knock them out." Jeremy's family is made of men of pretty good size. Jeremy, at this time, was 6'1" and weighing somewhere around 190, all pure muscle. He wasn't a little guy, but in his eyes, his brother was huge.

They were about the same height but there was no size comparison. When drunken incidents happened, Jeremy's brother would handle it, and if they didn't settle down or came at him, he just knocked them down. This was how Jeremy learned to handle disputes. At seventeen and full of young man testosterone, he stepped into this role easily and eagerly. When fights and disagreements happened, Jeremy would get right in the middle of them, resolving them in the same manner as his brother.

He left the carnival early that summer to attend a football camp at the University of Arkansas and to work with the coaching staff of Houston Nutt. Jeremy thought he was a pretty good sized individual until he got to the camp and got to see the former players from the program. He realized he had some growing to do. He did well enough in the camp and it grew his desire to play football in college. Coming back home from that camp, he and the group of players going into their senior year really bonded and came together. Jeremy brought the aggression he had acquired working with the carnival to the football field. Coach May, his position coach, began to teach him how to channel this aggression into becoming a stronger leader of the team. Coach May wasn't the yelling, screaming type of coach but could get the most out of Jeremy on the field.

However, this aggression also showed itself in school. The high school principal recognized this early and he sought to get a handle on it. The first week began and Jeremy continued right on with his aggression. He was a senior, a football player, and leader in his class. He was loud and obnoxious and students were following his behavior. The principal knew he had to bring this to a halt and called Jeremy and a couple of other boys into his office and informed them they were getting licks due to their behavior.

He administered the "licks" – swats with a wooden paddle - and asked Jeremy what he thought about that. Jeremy's response was, "It didn't hurt," resulting in a couple more licks before they were done. This resulted in a very contentious relationship between Jeremy and the principal for the remainder of the year.

Also at that time, he became enamored of an exchange student from Sweden. She became Jeremy's girlfriend shortly thereafter, primarily because Beth would still not date him. This relationship lasted all through his senior year and was different on so many levels. Jeremy came from a conservative background and she came from a European background of sexual freedom and independence. This turned out to be a dangerous combination for a seventeen-year-old boy full of testosterone and a seventeen-year-old young woman with no boundaries. Add to this that Jeremy had his own unsupervised house and, as Jeremy himself stated, "We made a lot of stupid decisions that year, which could have drastically changed both our lives forever."

Football season began and Greenbrier won their first game rather handily. After coming off of a losing season, this was a good thing. They also won their second game in the same manner. Now Jeremy not only has a girlfriend and the life all high school seniors want, his football team is now experiencing success. People began to take notice. In their third game, the team they were playing had a very good running back and he was giving them trouble. Jeremy took matters into his own hands during a tackle. The running back was hit at the line of scrimmage and his legs flipped in Jeremy's direction. Jeremy took the back's ankle and turned it as the linemen drove the player to the ground. This resulted in the running back going out for the remainder of the game.

Coach May pulled Jeremy out immediately and told him in no uncertain terms that this type of behavior would not be tolerated. Playing hard and playing dirty were two different things, and playing dirty was not an option. His coach cared about him and knew this was more of a moment of opportunity that was poorly executed rather than a common thing. Thankfully, the coach seized the moment to impart wisdom unto Jeremy. This weighed heavily on his mind and affected how he played the remainder of the year.

They won the third game, and this losing team from a year ago was now 3-0. Conference play started and things changed. They lost their first conference game, which also happened to be homecoming, in overtime. Jeremy remembered this was the first time he could recall not having anything left to give after a game and it not being enough. This conference was one of the toughest at this classification level in the state. In fact, the two finalists in the state playoffs that year for this classification came from their conference. There were no easy games, and the team went on a four game losing streak.

They were competitive in most of the games but couldn't find a way to get over the top. However, as the season came to its end, beating their school rival would give them a respectable record and also give them bragging rights. As Jeremy stated, "We could be one and nine as long as that one win came against Vilonia." The game was at home and Jeremy's team came out on top, putting together as close to a perfect game as they could. It was a phenomenal end to a great season for Greenbrier. Jeremy again garnered All-Conference honors, and he was contacted by Ouachita Baptist University about playing at the next level. Unfortunately, he didn't receive any scholarship offers but

OBU did give him the chance to walk on and the opportunity to be part of their team.

With football over, Jeremy's interests changed. His relationship with Beth was still of a friendship type, but even so, they became closer as their senior year wore on. He and Beth would go on double dates, Jeremy with his girlfriend and Beth with a boyfriend. They had to sneak a lot of those because Beth's mom didn't want her to have any part of the boy she was dating. Jeremy, in the far back of his mind, always knew the relationship with his girlfriend was a temporary thing. Even though they dated all through the remainder of the school year, it always seemed awkward for Jeremy. There were so many cultural differences between them and he struggled trying to reconcile her European ways to his small town Arkansas ways. But things were going pretty good and things were happening he didn't want to stop, so he continued in this relationship.

Grades were not that important, but Jeremy managed to maintain a good enough GPA to get him through school. He messed around with several other extracurricular activities, including parliamentary speaking and theater. But as always, especially now that he was going to college to play football, Jeremy never lost focus on preparing for the next level. He also continued to excel in his lack of respect for authority.

Near the end of their last high school days, Jeremy organized a couple of activities that went against the administration of the school. The first of these revolved around what was commonly known as Senior Skip Day. This is a day when all of the seniors skip class for a day. Jeremy took it a step farther in that he wanted all of the football seniors and whoever else to spend the night on the football field. Needless to say, this didn't go over very well and everybody who participated was forced to leave the field early before the administrators showed up. A large number of them went to Jeremy's house to spend the remainder of Skip Day. The principal found out and called Jeremy's house. Jeremy's mom happened to be home and took the call. He demanded the students all return to school for punishment, but Jeremy's mom basically backed him down from his demands and Skip Day continued.

The seniors also had shirts printed up that frankly were not appropriate for wear to a school. The shirt had a beer keg on the back with beer dripping from the tap and a logo that read "Seniors 2000 tapped out." Most all of the seniors wore these shirts on their next day back to school. The principal grabbed the ringleaders, especially Jeremy. He told all of them there wouldn't be any punishment as long as they replaced their shirts immediately. They left his office and since Jeremy and a couple of the others didn't have any other shirts to wear, they asked one of the teachers if they could just turn them around. Of course, all this did was really highlight the keg design. The principal saw this and became enraged. He called the three of them back to his office and told them to clean out their lockers and they wouldn't be allowed to walk the graduation line with the rest of their classmates. Jeremy cleaned out his locker and was sent home. Fortunately, common sense took over for both sides, and all three young men were allowed to graduate with their class.

The summer brought about the first of many changes in Jeremy's life. The initial change was that Jeremy did not go on the road with the carnival after school was finished. Instead, he followed his girlfriend to Sweden. This wasn't a permanent thing because he knew he needed to be back by the end of July to begin preparing to play college football. Even though he had traveled around the country with the carnival, Jeremy experienced a totally different culture in Sweden. Everything was different. She had to attend her school in the summer, and that allowed Jeremy to basically experience the culture on his own. Upon arrival back in the States, seeds of doubt about this relationship began to grow large. He began to see this relationship was probably not going to pass the test of time and distance, let alone the differences in both of their values.

Everything Jeremy was doing was outside of God's will for his life. Jeremy continued to attend church and his youth group and play the part of the good Christian. Beth continued to be a positive influence on his life but all of his activities within the church were basically play acting. As Jeremy continued to move through life on his own terms, God never quit pursuing him. He constantly found himself battling between his own free will and what the Spirit of God was telling him was wrong. Every time Jeremy seemed about to step over the line, God was there telling him he needed to back up and look at what he was doing. It was also at this time that God began removing things from Jeremy's life that Jeremy had turned into false idols.

Chapter 7

Removal of Idols

To say Jeremy wasn't an outstanding student in high school might be a bit of an understatement. His focus was never on getting good grades but in getting good enough grades to get by. His primary focus was on football. He knew he needed a certain GPA to continue to play and he maintained that GPA. He also loved all of the extra activities (girls and partying) that came with being in high school, and making good grades just wasn't a high priority.

Jeremy's acceptance to OBU to play football also came with his being placed on academic probation immediately. In fact, if it weren't for football, it was questionable whether or not he would have been accepted. Upon returning from Sweden, Jeremy had less than a month to prepare to report to training camp. His mom had stayed home from the carnival to help him prepare to leave. They got clothes together and items they thought he would need. He also received the workout program all reporting athletes are given prior to reporting to camp. The coaches want players to be in some semblance of readiness so they give them a workout program to do during the weeks prior to reporting. Jeremy attacked this program like he had attacked all of the others. This program was designed to add bulk to Jeremy's frame, and it did just that. He put on at least fifteen pounds of muscle bulk doing exactly what the workout program called for. This moved him over 200 pounds for the first time in his life.

Throughout this time, God was continuing to work on Jeremy's heart. Doubt about his relationship was continuing to grow and the question of whether he was on the right path for his life began to surface. In a way that only God can work, questioning one thing in Jeremy's life caused him to begin to question everything in his life. Jeremy didn't realize it at this time but God had begun the process of removing all of the idols from his life. It took a bit of time but this was the beginning of the process.

Jeremy packed everything he basically owned into two duffels, and his mom drove him to OBU to report for training camp. Jeremy immediately experienced his second bout with cultural shock in less than four months. Instead of a locker room full of his buddies from a predominantly white, rural school, he was thrust into a locker room that was culturally and racially diverse. He was an incoming walk-on freshman, the lowest of the low on the totem pole. Not only was the locker room diverse, he was now encountering twenty-one- and twenty-two-year old young men who have fully developed, mature physiques. He was a stud in Greenbrier but here, he was one of the smaller guys. It's not that he was intimidated, but overwhelmed would be a better description.

Two-a-day practices began and they were quite different than his high school experience. Within just a few days, Jeremy found himself being pummeled on a regular basis by these bigger, faster, more experienced players. It was a tough experience for young man who was used to being cock of the walk on his team and his school. The doubt about not being on the right path for his life began to manifest itself. By the end of the first week, the thought of giving up football was forefront in his thoughts. Jeremy said, "I don't know that this was the right decision any more. Not that I was getting my head kicked in every day. That definitely didn't help the situation, but I really felt like I was not in the right place." When he said this to his girlfriend, she was taken aback. She liked the idea of having a college football boyfriend, and this poked another hole in their relationship.

While all of this was going on, the first idol God took away was all of Jeremy's friendships. Jeremy, at this point in his life, was now all alone. Most of his friends stayed in Greenbrier, with the exception of a couple of guys who went to Henderson State University, the crosstown rival, to study aviation and become pilots. But none of those individuals were now in Jeremy's life. He had nowhere and nobody to turn to or talk to. For an eighteen-year-old young man, this can be extremely troubling. His buddies weren't there anymore. He was without the presence of his girlfriend for the first time in over a year. He found himself with nowhere to turn, and God used this time to weigh heavily on Jeremy's heart.

Football, to the shock of everyone who knew him, became the second idol to fall. This was the hardest decision Jeremy had ever had to make in his young life. Football had been his number one priority and focus as long as he could remember. He frankly couldn't remember very many times when his life did not revolve around playing football. He had never quit anything in his entire life. "I was an absolute train wreck, all over the map," was Jeremy's description of himself at this time. He was angry at himself and felt like he was letting everybody down. He didn't have any direction or guidance, and he was scared to death to walk into the coach's office and tell him this decision. This felt like failure and he didn't understand it since he had never before failed at anything. Jeremy finally worked up the courage to tell Coach Knight of his decision. He still didn't understand that God was leading in this decision. He just knew in his heart he was not in the place he was supposed to be and sensed that God was leading in this direction.

The day Jeremy told Coach Knight was at the same time the most lost and the most relieved he had been. Jeremy was in tears as he told Coach Knight he was quitting the team. "I don't know what to do, but I know I'm not supposed to stay here," started the discussion. Jeremy never alluded to anything about God leading him away from football but was amazed that Coach Knight responded with grace and mercy. Coach Knight quit being a football coach and became a counselor to Jeremy. After Jeremy said everything he had to say, Coach Knight said, "You are still in school. You are here to get an education. You are not here to play football. If that is not the direction God has set out for you, then you do not need to play football and let that be in the way. If He has taken that away from you, it is OK."

Coach Knight's discussion with Jeremy showed him a path. He thought because he was quitting football meant he was quitting school, but Coach Knight quickly disposed of that thought. He allowed Jeremy to stay in the football dorm until school started and he could obtain different living arrangements. Before Jeremy left his office, Coach Knight prayed with him, and when Jeremy left the office, he knew he was at peace.

Dealing with diverse individuals continued as Jeremy was moved from the football dorm to the student dorm. He acquired three new roommates. The living arrangements were in what they call a pod unit. It consisted of three rooms, with the bathroom being between the two major living areas. Two people slept in one area and two people slept in the other, with all four sharing the bathroom. Jeremy's roommate was the polar opposite of the types of individuals he normally ran with. The roommate, in football jock vernacular, was a geek. Trouble started immediately when the roommate hung a Pokémon poster on the outside of their door. Jeremy demanded the poster come down because he did not want people thinking he was a geek. That was not his life. The next issue arose when trying to decide placement of bunks. The roommate did not want the bunks stacked but rather wanted them side by side. Jeremy's resolution to this was to put end tables between the two bunks. Needless to say, it didn't get any better as the semester progressed, but this actually helped lead Jeremy closer to God.

Jeremy's method of getting out of his room was to go downstairs and visit with a couple of guys. One was an acquaintance from Greenbrier, Bryant Turney, and his roommate, Chris Daugherty. Jeremy began hanging out with Bryant and Chris. They would study, eat, and socialize together. Bryant invited Jeremy to church and he began to go to Cornerstone Baptist Church in Arkadelphia, Arkansas. Bryant and Chris were true believers, and they were the start of the new path God had in mind for Jeremy. Not only was Jeremy hanging around with Bryant and Chris, he also began socializing with their friends. The influences on his life were changing drastically.

In late September, Jeremy went to a party at Henderson with his old friends and had too much to drink. He staggered back to his dorm room and fell asleep on his bed. The next morning, one of the roommates from the other room came to check on him. He asked Jeremy if he was OK because he had been so sick the night before. Jeremy responded he was fine and as the roommate turned to leave, he looked at Jeremy and said, "You need Jesus."

When he said that, Jeremy felt the Holy Spirit say to him, "I'm changing your heart. Your desires are now my desires. Your life is going to be marked by me." Jeremy accepted Jesus as his Savior and his life changed at that very moment. There was no more pretend, there was no more acting. Jeremy became a child of God, knowing that Jesus died for his sins, and asked God to come into his life.

Jeremy immediately went to Bryant and Chris's room, told them what had transpired, and started asking for direction. They began praying and counseling with him and started helping Jeremy in his baby walk with God. They shared their salvation stories with Jeremy and helped confirm with him that his salvation was real. The first weekend after his salvation, Jeremy went home and visited with his old pastor. Sadly, instead of embracing Jeremy's salvation, his pastor treated it as a recommitment. Jeremy also had the chance to share his news with Beth, and she was overjoyed because she knew the difference. From this point in time, Jeremy's desire for his old life was gone. Beth was the only person God left in place. Jeremy thought about Exodus 6:7: "***I will take you as my own people, and I will be your God.***" (NIV) After the first week, Bryant and Chris made sure Jeremy started attending a small group Bible study.

OBU had a program called "Refuge." This was a worship service in addition to the chapel service offered each week. In addition to chapel and his small group, Jeremy started attending Refuge. It was led by Michael Bleeker. These worship services served to open Jeremy's exposure to Jesus and broke down barriers that had been placed by traditional, fundamentalist churches. He was amazed to see worshipers raising their hands and moving to the music and became a regular at these services.

In October, at fall break, Jeremy went back to Sweden for a visit. God, at that point, began removing the last of Jeremy's idols. In Jeremy's words, "It was the coldest, weirdest week I had ever encountered." Everything he tried to do felt awkward and stressed. It didn't feel the same. His girlfriend was indifferent to Jeremy's revelation because it was not that big of a deal in her culture. Even though the two did not separate at this time, there was now a chasm in their relationship.

The one friend from home Jeremy kept in contact with was Beth. She did not get to OBU the first semester, instead going to UCA, which was less than twenty miles from Greenbrier. She was able to live at home and commute, allowing her to begin saving in order to attend OBU. Jeremy, Bryant, and others would travel together back to Greenbrier in order to be able to travel home and share fuel costs. One of those was a guy named Andy, and he had become a pretty solid part of their group. On one trip home, Jeremy introduced Andy to Beth, and it wasn't long after that Andy and Beth began dating.

During Christmas break, Jeremy's parents, decide they need to travel to Sweden and meet his girlfriend's family. Jeremy believed in his heart his parents hated the trip. First, it's Sweden in the winter. Cold doesn't begin to describe the weather. They toured all of the sights and when they left, Jeremy began to seriously understand his relationship with this girl was headed nowhere. His growth as a Christian was creating a barrier to the behaviors the couple had experienced early in their relationship. His family returned to Arkansas just before Christmas and to the one constant relationship in Jeremy's life. Beth was there virtually every day. Beth informed him she had been accepted and was enrolling at OBU for the spring semester, and Jeremy helped her move from Greenbrier to the student dorms in January.

At the start of the spring semester, OBU's "social clubs" began accepting pledges. Jeremy, Bryant, and Chris started attending get-togethers at the social clubs, and they pledged

Chi Iota Sigma, the letters representing compassion, integrity, and self-discipline. It was the smallest club on campus. When he told his girlfriend about it, she didn't want him belonging to a Christian group and instead wanted him to be part of a "cool" group. This pretty much sealed the deal.

This led to his being exposed to serious discipleship and Bible study the likes of which he had never experienced. This club consisted of less than twenty members and they were extremely serious about their relationships with Christ and expanding those relationships. They moved Jeremy into a leadership position of leading a Disciple Now (Dnow) event in a small city close to Little Rock. He didn't think he was ready for this because his knowledge of the Bible was limited, but because he had the ability to speak in front of crowds, he was pushed into this leadership role. What he didn't understand at the time was that this was part of the process teaching him to respond and grow. He had to study and he had to understand what he was talking about, and being pushed in this manner accomplished that goal. This became a weekly event, leading Dnow events at various churches throughout the state. This actually created more questions from him about what he was hearing from other speakers and pushed him to find answers to those questions. This in turn enabled him to help high school students when they came to him with these same questions.

In March, at spring break, his girlfriend came to the U.S. for a visit, and the couple went to Florida. Jeremy knew the relationship wasn't right, and it quickly became evident during this week. They drove back to Arkansas in near silence the entire trip. Upon arrival back at the airport, things were clearly coming to an end. She got on the plane and went home. Jeremy realized, as he watched her walk down the jetway to her flight, that God had removed the last earthly idol in his life. A very short time later, a long distance phone call finally ended it all. Again, Jeremy had a mixture of emotions. This was a multi-year relationship with someone with whom, at one point, he thought he was going to spend the rest of his life. But he also knew that God was in the process of changing him, and deep in his heart, Jeremy knew this wasn't the life-mate God intended for him. Upon arrival to OBU, his friends gave him the best advice he could get. They told him he just needed to be single for a while and let God work with him.

The removal of all of the idols in Jeremy's life brought him to a total dependence on God. Even though he tried, he was not able to "fix" any of the problems with these idols. He discovered what all true believers experience at one point or another: God does not want you to fix these problems. He wants us to lean on Him and have total dependence on Him and what He can give us. ***"Trust in the Lord with all your heart and lean not on your own understanding"*** Proverbs 3:5 (NIV). This is when Jeremy quit depending upon himself and began depending and trusting upon God.

Jeremy's grades in his first semester were just like they were in high school, and the class that allowed him to stay in school was scuba diving. His "A" in that class made his GPA high enough for him to register for the spring semester. Because God had removed all of the earthly idols Jeremy had worshiped in his earlier life, he now found he had time to spend studying and learning about the Scriptures and the Lord to whom he had surrendered his life. With football, his high school friends, and relationships gone, he also had time to seriously devote to his studies. His plan, from the beginning, was to major in pre-law, go to law school, and become a lawyer. From making barely passing grades, he began making A's and B's in all of his classes. The stability that God had now instilled in his life was beginning to bear fruit. He became a member of the debate team that throughout his college career garnered several national honors, and he never again struggled in school.

Through the rest of the school year and into the summer, Jeremy counseled with and was mentored by the guys in his club and also by the youth pastor at his home church in Greenbrier. He clung to this mentorship and reacted when they called out previous things in his life. Unfortunately, becoming a Christian and a true believer doesn't mean he was instantly endowed with all of the Biblical knowledge and experiences of a mature Christian. He was like a babe learning how to crawl and walk. These guys helped Jeremy as he learned more and more about his Savior and the Scriptures.

Chapter 8

Beth

While all of this was happening, Beth continued to date Andy. It seemed as if every time Jeremy was single, Beth was in a relationship. It didn't diminish the feelings that were growing inside Jeremy concerning her, but he was not in a position to act upon them. However, any chance he got to be around her, he took. They would travel back and forth to Greenbrier for family visits on the weekends, spending hours in the car just talking as they traveled. Beth also began to see changes in Jeremy's life, realizing he was not the same person he was in high school.

Beth and Andy dated through the summer as Jeremy traveled with his family on the carnival trail. Andy went on a mission trip to Africa, and his contact with Beth was limited. Even though Jeremy was traveling, he still made time to talk to Beth. She began calling him as much as he was calling her. The friendship that began to blossom during the school year continued to grow.

One day, Beth called Jeremy and told him about her excitement over Andy's coming home from his mission trip. She was driving to his parents' home in Mississippi to meet and surprise him. Obviously, Andy was surprised because after meeting him at the airport and going back to his parent's home, Andy broke up with her. His reasoning that was he was going to go into the mission field and he wanted a wife who had the same calling. Other than that, there was no reason given. Beth was devastated. She called Jeremy while she was driving home that evening and told him about the breakup. Jeremy had two distinct thoughts: 1) that guy is a moron, and 2) how fast can I get to her.

The next morning, Jeremy called Andy and asked him why he left her. The conversation was pretty brief because once Andy told Jeremy he didn't want to be in a relationship any longer, Jeremy realized he didn't care about listening to anything else. As soon as he got off of the phone, he told his parents he was leaving and drove to Beth's house. When he got to her parents' home, she wasn't there. He had bought some flowers and asked if he could leave the flowers and a note. The note expressed Jeremy's heart; he was absolutely in love with her and had been for a while. The note continued that she was always on his mind and he knew they were meant to be together. He left the flowers and note on her bed and he left. As Jeremy looked back on it, he laughed and recalled his timing could not have been worse. Beth came home, saw the flowers and note, and called Jeremy about meeting. Jeremy was ecstatic because she wanted to meet right away. But what she had to say was not what he wanted to hear. She told him she had just gotten out of a relationship and was not ready for another right away. She wanted to be friends, and this totally deflated his balloon. She loved the support he was giving her but that was all she wanted it to be.

The remainder of the summer, they continued to be friends and socialize, and now Jeremy found himself keeping potential suitors away from her. Jeremy knew God had changed him and also knew that God had not removed Jeremy's desire for her. This was another lesson God was teaching Jeremy. He distinctly remembers in his mind that he needed to give it time and God would open the door to Beth's heart when He saw fit. In every Christian's life there comes a time when one must wait upon God. Sometimes we have to wait on God for fulfillment of a specific desire. It's how God teaches us patience, and that was what God was teaching Jeremy at this time.

Upon returning to OBU in the fall, things were different in a lot of ways. Jeremy and Beth were in school together and neither were dating. His old friends didn't have any pull or draw on him and new friends were mentoring him every day. Classes were a priority, and making good grades became the norm instead of the exception. He moved into a new room and a new living arrangement with his new friends. It was the first fall for as long as Jeremy could remember that football was not the priority in his life.

Through all of this, Beth and Jeremy were constantly together. They attended the same services, hung out together with the same group of friends, and, of course, continued their weekly trips back and forth to Greenbrier. Technically, they were not dating but they were also not seeing anybody else and were with each other at every opportunity. Things clicked along for everybody until September 11, 2001.

Chapter 9

9/11

That morning was a laid back morning. Jeremy was hanging around the dorm room, actually just relaxing in his bunk. There was chapel later in the morning and he was preparing to attend. Suddenly there was a loud banging on his door and shouting in the hallway. When Jeremy opened the door, the individual there said to him, "Are you not watching the TV? We are under attack!" Everybody gathered into one room and they stayed there watching everything unfold. Their phones began going off and messages were sent instructing all of the students to report to chapel.

Hundreds of people arrived at the chapel. Dr. Andy Westmoreland walked to the microphone and informed the students they would not be having chapel. Instead, this would be a time of extended prayer for our country and all that was happening. After an hour, they were all instructed to call their families, find their friends, and attend their small groups. Classes were cancelled for the day.

The family carnival was actually in Hot Springs, Arkansas, that day, which was only a 20- to 30-minute drive from campus. Jeremy's brother was managing that portion of the carnival and Jeremy drove to him. He talked to his parents by phone and was able to spend some quality time with his brother. On the drive back to campus, Jeremy came to the decision he would join the military to be able to exact revenge on those who dared to attack this country.

The next day he walked into the Armed Forces Recruiting facility. Captain Michael Washington was in the office that day, and as Jeremy walked in, he was instructed to take a form, fill it out, and bring it back. Jeremy started to argue with him because he wanted to join right then, but CPT Washington informed him he was about the fiftieth person to come in that day. Again, Jeremy was told to take the form, fill it out, and bring it back the next day.

Jeremy did just that. He took the stack of papers home with him and completed each one. He called his parents that evening to inform them of his decision and needless to say, they weren't quite as excited about this idea as Jeremy was. Both parents tried desperately to talk him out of doing this right away and wait to see what would happen. Like parents everywhere, they thought he was overreacting to the situation. But Jeremy would not be swayed, because he saw this as God clearly setting a path for him.

The next morning, Jeremy took his paperwork back to the recruiting office and left it because no one was in the office. Later that day, CPT Washington called asking him to come back to the office. Upon returning to the office, there were Jeremy and a young lady present. (This young lady currently serves in the Arkansas National Guard stationed at Camp Robinson with Jeremy.) Of the over fifty individuals who took packages the previous day, these were the only two people to bring them back.

CPT Washington, who happened to be in the Arkansas Army National Guard, did something very instrumental in both of their lives that day. Instead of Just enlisting them immediately into the Army and shipping them off to boot camp, he directed both of them into a program that would allow them to continue their education, attach them to a Guard unit as Guardsmen, and be able to obtain a commission upon graduation. Jeremy reported to the Military Entrance Processing Station (MEPS – where all processing for all military branches takes place) for his physical and induction and became a member of the Arkansas National Guard. He was moved into what is known as SMP, Simultaneous Membership Program. This program allows an enlistee to join a Guard unit, assume the rank of SMP cadet (paid as an E-5), and shadow an officer in a chosen career field until such time he/she graduates. He chose an artillery career field and from that time on, Jeremy was a National Guardsman with an artillery unit.

Chapter 10

Progress

In early October, there is an event at Ouachita called "Tiger Tunes." Essentially, it is a musical production by various organizations on campus to raise funds for a group called Students Helping Students. Jeremy's social club was part of this production, and Jeremy was one of the directors. Beth played a major part in their production, which meant they were now interacting in every aspect of their lives.

Shortly after the completion of "Tiger Tunes," Jeremy decided it was time to ask Beth on a date. A lot of time had passed and the time seemed right. She said she tentatively had plans but would definitely get back with him. It brought back a line from Dumb and Dumber, "So you say there's a chance." He didn't hear from her that day. Towards the end of that day, a buddy of his called and asked if he wanted to go to one of the comedy clubs. The friend also set him up with one of his female friends. As they were preparing to leave, Beth called and said she would go with him that night. Now he had a dilemma. He already had a date, but the woman of his dreams had now agreed to go out with him. Jeremy did the right thing, proving how much he had grown. The old Jeremy would have ditched his commitment and gone to Beth. But this was a new, follower of Christ, Jeremy. He was honest with her and told her the situation.

The date proved to be a disaster and reinforced even more his feelings for Beth. About two weeks later, Beth called Jeremy and asked to borrow twenty dollars. Jeremy was able to help her, they hugged and went their separate ways. Unbeknownst to Jeremy, Beth called her parents that evening and professed her love for Jeremy to them. After talking to her parents, she called Jeremy and said they should go out. Jeremy had no clue what she had told her parents. A couple of days later, they went on their first date, going to Hot Springs to a movie and hanging out for the evening. They finished the evening at Waffle House because it was the only thing open.

For the remainder of the year, everything fell into place. Grades were good, relationships were amazing, and Jeremy was gaining military experience with each monthly drill. He was learning what it took to not only be a good artillery officer but what it took to be a good member of the military. Each drill period brought about new experiences that he continues to draw on to this day. He was attentive to what his officers and troops were doing and saying. All of this fully prepared him for his first military period of training, boot camp at Fort Knox, Kentucky.

Chapter 11

Boot Camp

SMP cadets do a shorter, more condensed version of basic training. Even though he had been attending drills and some field training with his Guard unit, Jeremy really didn't know what to expect. Of course, having seen movies and listening to the guys in his unit, he had some sort of an idea what to expect at boot camp, but this would be different. He left Little Rock by plane and landed in Louisville, Kentucky. Nothing unusual about the airport experience. The military does not show its hand inside of the airport where it might be unsettling to the civilian population. The cadets gathered in one small location in the airport and were "asked" to get on the bus when it arrived. It was the last time he was "asked" to do anything for a while. Everything at this point appeared to be pretty normal, but things quickly changed upon arrival at Fort Knox, Kentucky.

The bus pulled up to a really ugly set of barracks and the drill sergeants arrived. They were like sharks circling wounded prey. Jeremy was given one piece of advice that is true in all boot camp situations. Do not do anything that will make you stand out. One of the drill sergeants was walking down the line asking questions and mocking the answers he was given. When he got to Jeremy, he asked him where he was from and Jeremy answered he was from Arkansas. He took a step back, hollered at the other drill sergeant and said, "We have one from Arkansas." Jeremy kind of shrunk because he was expecting some hillbilly joke. But instead, come to find out, both drill sergeants were from Arkansas. They worked him up one side and down the other, informing him they would be on him hard to make sure he did not fail and bring disgrace to the great state of Arkansas.

The rest of the evening was spent getting haircuts, uniforms, and shots. The drill sergeants were in their faces, constantly on them about something and everything. Jeremy also experienced his first ride in a "cattle car." When he got into the vehicle with his duffle bag and about twenty other people, he thought it was about full. Wrong thought. It was full about eighty people later.

It was hot. Not just hot, but southern hot. It was the end of June and summer was cranking up. They finally got to the barracks and began to get things squared away. Jeremy said, "When it was lights out, I got the best night's sleep I had had in a while. Unfortunately, it ended with the rudest awakening ever at about three in the morning." Both drill sergeants showed up yelling, screaming, and banging on everything. Welcome to the U.S. Army.

The next five weeks were spent becoming military indoctrinated. Proper marching was stressed in the first week, weapons were assigned and troops were taught to shoot properly. It was tough and at the end of each night, troops collapsed, only to wake up the next morning to the same thing, only a little more. After two weeks, they went on their first field exercise. It was a short, overnight exercise. To Jeremy, the funniest thing happened upon arrival from the field.

The drill sergeants gathered all of the male troops together and informed them they would have a mandatory tick check. This was done because in the south, the woods are full of ticks and chiggers, and they will find their way to bodies when given the opportunity. The troops were instructed to strip down, get in a line, go into the showers, and check themselves for ticks. They were told to ask the person behind to look at their back and remove any ticks that could not be reached. For people used to the woods, this was no big deal. For those from the city, this was a terrifying experience.

One particular soldier from a northern state who had never been in the woods let out a horrible scream. "I'VE GOT ONE ON ME!" The really bad part is that it had attached itself to his private parts. He kept yelling he didn't know what to do and was having a meltdown. This captured the attention of the drill sergeant and he came running. He told the cadet to show him and when he did, the drill sergeant pulled out his knife and said, "Looks like we are going to have to cut." The cadet turned white as a sheet and passed out. When the cadet gained consciousness the drill sergeant pulled out a pair of tweezers and handed them to the cadet who commenced getting the tick by himself. Knives can be great motivators.

The cadets settled into a routine. Sundays were a time to go to chapel, and everybody went to chapel. This was Jeremy's first exposure to military chaplaincy. Didn't make much of an impression, but Jeremy was able to find time to get into the Scriptures when he pulled CQ (Charge of Quarters) duty. He would make his rounds and do his checks and was then able to sit down and read his Bible. It was an hour every night to do what he wanted to do. The only true day off came on July 4th when there were no duties assigned. This was also the day Jeremy realized he was exactly where he was supposed to be. God gave him a peace in his heart that he was on the path God had chosen for him. ***"I've said these things to you so that you may have peace in me. In the world you have distress. But be encouraged. I have conquered the world."*** John 16:33 (NIV)

The mentorship Jeremy had prior to arriving at Fort Knox paid off. Jeremy embraced the concept that he would be leading troops in a couple of years, and he bought into this concept. He understood the fact that he would one day be a leader. Whenever he was put into a leadership position, be it either the platoon leader or platoon sergeant, he treated the platoon like it was his platoon and his total responsibility. He also understood when he wasn't in a leadership role, it was his responsibility to do his absolute best for the individual in charge. This mindset was ingrained in him when it was time to graduate and he took it back to his unit.

Chapter 12
She said "YES"

When Jeremy returned from Fort Knox, he went back to work on the carnival for the remainder of the summer. It would prove to be his last summer there. At the end of summer, Jeremy and Beth returned for their junior year at OBU. Grades continued to improve and the commanders of the ROTC unit moved Jeremy into a leadership position, a direct result of his embracing the leadership concept at Fort Knox. The whole year kind of became a blur, with nothing important happening except for two events. One was that Jeremy was selected for Mountain Warfare School, becoming the first cadet from the state of Arkansas ever selected for this school. The reason he had applied went back to his reason for joining the Army in the first place. He wanted revenge, and he knew Afghanistan had mountains.

The second event, and most important, was Jeremy's planned marriage proposal to Beth. There really wasn't any surprise because Beth was ready for this. However, during spring break, while home in Greenbrier, Jeremy planned a competition date night. They played a multitude of games that day, things like pool, bowling, and horseshoes. The winner got to express a wish and the other had to fulfill it, and Jeremy thought this was a win-win situation. At the end of the night, the competition was tied, with a game of pool determining the winner. Jeremy lost, some might say on purpose. Beth looked at him and said, "I wish you would propose to me." They sat down on the couch, Jeremy got down on one knee in front of her, asked her to marry him, and gave her the ring. Wasn't exactly according to Jeremy's plan, but in his words, "It was a perfect day and culminated in the perfect evening."

What surprised Jeremy was how fast and how elaborately a wedding could be planned. He proposed in April, and the following January, before returning for the final semester of college, they were married. Of course, like most young women, Beth already knew how she wanted her wedding to go. Jeremy spent as much time as he could looking for a house, and he found a small, blue house that was a little run down, situated between two OBU professors' homes. It was what they could afford.

This eight-month period between the proposal and the wedding was spent fine-tuning his military duties, including going to two different military schools that summer, growing closer to God through his studies, getting the fall semester of his senior year in line, and getting a job.

He found a job at Twin Rivers Nursing Home and was hired as a nursing assistant, working for minimum wage on the night shift. His strength came in handy since he was required to lift and move patients. The administrators also discovered he was moving into the ministry and asked him if he would want to minister to the patients. He was able to pray with the individuals in the home and ministered to all of the patients on his hall. He continued to work at this job until he graduated, eventually moving into a supervisory role.

January rolled around and all they had worked for the previous eight months came to be. Jeremy thought he was as ready as he could be. He had a job and a house, things were going well in school, and in the military part of his life, all was well. He remembers the wedding ceremony being the longest ceremony ever. For a Baptist wedding, its length was more representative of a Roman Catholic wedding. The preamble to the wedding took forty minutes due to the number of people singing and groups playing. Once Jeremy and Beth took center stage, things moved more quickly, and a fireworks display topped off the evening. They honeymooned in the Poconos and returned immediately to begin the spring semester. Jeremy had already moved into the house, and Beth moved in when they returned from their honeymoon.

Chapter 13

Advanced Camp

At the end of the semester, Jeremy attended Advanced Camp in Ft. Lewis, Washington, a five-week course whose purpose was to refine leadership skills. The terrain reminded Jeremy of the Star Wars movie that introduced the Ewoks and where they lived. Every platoon lived in WWII barracks. Each platoon was placed in different scenarios, and each member of the platoon was placed in a leadership position during this time to judge their performance. It was emphasized to the cadets that they would always be under evaluation, whether as a leader or as a member of the platoon. Individuals were rotated out of leadership positions in the platoon with one caveat: If a cadet was not doing well, they would be replaced. However, if they were doing well, they would be left in place to see how far they could go. This course could literally determine if a ROTC cadet would be able to continue in their career. How individuals finished in this course also determined how they would be considered for assignment upon graduation. It also would determine which cadets needed to be removed from the ROTC program, and for many cadets, this would be their last experience with ROTC and the Army.

About a week and a half into the course, Jeremy was placed in the platoon officer position and another individual was placed into the platoon sergeant position. They went through the day and went to bed that night knowing they hadn't made any mistakes. They got up early the next morning for breakfast, preparing for their replacements to show up. However, the instructor arrived and told them they were still in charge. They stayed in these positions for a couple of weeks, leading the platoon into their first battle drill. They got the platoon into the field and to their assigned positions, getting them bedded down and establishing a perimeter. At this point, the instructor came to both of them and advised them they were being relieved. He explained to both cadets they had done nothing wrong, it was the last opportunity to put somebody else in charge for evaluation.

Later that evening, the instructor approached Jeremy and had an at length discussion about why he had replaced them, noting it had nothing to do with their performance. The instructor asked Jeremy what he wanted to do in the Army, and Jeremy mentioned he thought God may be leading him into the ministry. Unbeknownst to Jeremy, his instructor was a born-again Christian, and he told Jeremy his leadership style resembled the style of a pastor. The instructor said he not only led his troops but guided them spiritually throughout the past two-plus weeks and thought this was the reason they had performed so well.

On their last night in the field, they were embedded on a hill and Jeremy was bunkered with an individual of Mormon faith. This guy knew Jeremy was a Christian and began asking questions about his faith. It was a rainy, cold night, and they were shoulder to shoulder under a poncho trying to stay warm. Yet here they were, having an in-depth discussion about his faith. It was at this point the idea of Jeremy becoming a chaplain began to grow. This was the spot God used to pave the way for Jeremy's life journey.

At the end of the course, Jeremy was ranked number one in his platoon and was evaluated against the other number one individuals from the other nine platoons. The colonel doing the evaluations questioned each individual, and at the end of the evening, presented his rankings. Rankings one through three all went to individuals immediately finished with ROTC and headed to active duty postings. Jeremy was 4th and the top-ranked cadet headed back to college.

Upon return from this school, Jeremy had a week before he had to leave again for Mountain Warfare School with the 10th Mountain Division in Vermont. This was baptism by fire for Beth because she had never experienced anything like this before. What she was starting to experience was the life of a military spouse.

The school helped train troops on how to move quickly through mountainous terrain in order to help lead troops on the battlefield of Afghanistan. Each day, the trainees would pack their rucks full of ropes and other needed gear before road marching and climbing all day. They were tested on their ability to build rope systems, mountain land navigation, and medical evacuation of personnel on dangerous terrain. More than once, Jeremy found himself pushed to the physical limit. More people left the course with foot injuries than he had ever seen. Each night, the trainees would help patch up each other's feet to safeguard against the blisters. Moleskin (a foam patch) and duct tape became every person's best friend.

Jeremy was near the end of this school and found himself reaching muscle failure on the last land navigation course of the school. He lay on the rocks after climbing to his final check point to click the code into his punch card. After what seemed like an hour, he lowered himself down the cliff face and staggered to the headquarters to turn in his score card. Outside the building he watched as numerous trainees were extremely angry at having failed the course on this final day of testing. Jeremy handed his card to instructors to check the codes and awaited with dread the potential of terrible news.

Instead, they congratulated him on making graduation, and told him to check with the medics before grabbing a meal. His feeling of receiving terrible news was just moments away, though.

Upon reaching the barracks, he received an emergency phone call from Beth. Jeremy's dad had a massive heart attack and had died several times on the table. This hit Jeremy hard, not only because it was his dad but also because he knew his dad was not a Christian. The chaplain for the command sat and prayed with Jeremy, making a huge impression on him, and further helped grow the seed of Jeremy's becoming a chaplain. The school got his bags packed, informed him he had done enough to graduate, and got him to the airport for his flight home. Being the military, they flew Jeremy to Little Rock, because this was the nearest airport to his home. However, the heart attack happened at a carnival in St. Louis so upon arriving in Little Rock, Beth met him at the airport with good news that his dad was stable now and resting. They collected his luggage and boarded another flight to St. Louis. They spent several days with his dad, watching him improve every day. If there was a bright side to this experience, it's that it scared Jeremy's dad, and God used this as an opportunity to plant the seed of salvation, which would later come.

Chapter 14

Watching God Work

Upon returning to his ROTC unit, Jeremy was notified of his "branching." This means he found out in what career field the Army placed him. At times, the Army does make some pretty logical decisions. Since Jeremy was already assigned to and working in an artillery unit, he was branched as an artillery officer. At this time, Jeremy informed them he wanted to go into the chaplaincy. The decision was made in his head and his heart. He was informed it didn't work that way. He must first be commissioned into a designated field and then begin the process of applying for and being accepted into the Chaplaincy Corp.

Something happened that summer that cemented in his mind God was calling him into the ministry and to become a chaplain. Jeremy's home church was in need of a youth pastor and they contacted Jeremy, asking him to take over their youth group as a volunteer. Without hesitation Jeremy accepted the offer.

The reason there was no hesitation was because Jeremy had already begun these discussions with God as his dad lay in the hospital recovering from his heart attack. Jeremy realized his dad was not a believer, and he knew if he died he would spend eternity in hell. God used this point in Jeremy's life to dig deeper into his heart in regards to where God was leading him. He had been deeply impressed with how the chaplain in Vermont had worked with and guided him through the process with his dad. Jeremy began to lean heavily on God to help guide him through this time and also through the processes beginning in his life. He was newly married, would be commissioned in a year, and was in danger of losing his dad. God used that time to bring Jeremy closer to Him and set the path along which he would follow.

One special event took place with Jeremy while his father was recovering in the hospital in St. Louis. There was a small chapel in the hospital, and one day, Jeremy spent a couple of hours there in prayer. During this time of prayer, God revealed some things to Jeremy and helped him begin to understand his calling. When he left that chapel, Jeremy knew he was going to trust God, follow Him, but not know what the next step or the next journey would be. The only thing he knew was he was now following God's path and God had given him a peace in his heart for the rest of his endeavors.

Jeremy and Beth began driving every Friday night to Greenbrier, to do a youth event on Saturday night, attend church on Sunday morning and evening, and drive back to Ouachita late Sunday night. This was an intense ministry learning curve and there were a lot of failures. Jeremy was 21 years of age and leading teenagers who were not much younger than he was. He leaned heavily on the pastor, on books that were left behind in the youth pastor's office, and on the small Christian bookstore in Arkadelphia. This was Jeremy's life through most of his senior year in college. One Saturday night shortly before Jeremy was due to graduate, he was getting ready to lead the youth and the pastor came to him. He explained to Jeremy they were bringing in a more experienced youth pastor. He got two lessons from this. First, as quickly as doors open, they can close. The second was it gave Jeremy a taste of what ministry was going to look like.

Jeremy was neither licensed nor ordained during this year, but it didn't matter. He graduated and was to be commissioned as an artillery officer. This was also when God did the neatest thing. Jeremy had planned and was ready to spend several years as an artillery officer before beginning his journey to becoming a chaplain. It's always interesting to make plans when God is involved because very rarely do our plans look like His.

In 2003, the 39th Infantry Brigade of the Arkansas National Guard, of which Jeremy's unit is a member, was notified they would be mobilizing for a year in Iraq, and Jeremy realized he would be going to war. However, he was informed that because he had not yet been commissioned and would not be able to get commissioned in time, he would not be on this deployment. Some of the cadets who were closer to graduation were able to switch their branches in order to make the deadline for deployments. They all wanted to go, but only a few would be able to pull off the timeline. It hurt him to his core to see the unit he trained with head to a war zone and he was not able to go. Jeremy was also informed he would not be attending the field artillery officer course for at least eight months due to a huge backlog in students.

Shortly thereafter, he got a phone call from Dr. Tom South, Dean of Baptist Theological Seminary in Conway, Arkansas. He told Jeremy he was looking at his application to come to seminary and had a few questions. This puzzled Jeremy because he could not remember filling out an application to attend seminary. He had no idea how Dr. South had gotten an application from him. He was shocked to get the call and even more shocked to learn he had been tentatively accepted into seminary after getting them a few other items to finish the application. That same day, he also got a phone call informing him the National Guard was going to get him on track to become a chaplain. They knew they could produce artillery officers but it was much harder and took much longer to produce a chaplain.

As planned, the Arkansas Army National Guard was activated and deployed to Iraq in 2004 and Jeremy did not go with them. He stood with his troops in Morrilton, Arkansas, at the armory where Alpha Battery, 1st of the 206th Field Artillery calls home. He shook hands with everyone and loaded bags while they hugged their families goodbye. A short time later, the buses bound for Ft. Hood rolled out of sight, and it was Jeremy along with a single sergeant left to close up the building. For many, seeing their friends and unit leave can be tough. You feel like you are missing out on an opportunity or not able to fulfill an obligation. However, after receiving both of his earlier phone calls, Jeremy knew he was going to seminary but did not know what his military service would be during this time of deployment and waiting for the schools to open up. Once again, it wasn't a problem, because God had already worked it out.

The next day, he reported for annual training and was placed with people who would help him become a chaplain. He applied for and was hired for an active duty position at Camp Robinson for a year's period of time. They allowed him to set his work schedule around his seminary classes. During the day, he supported the logistical needs of the unit, and at night, he would attend school.

To recap, after graduation, Jeremy was commissioned and waiting for an artillery school spot to open up. He had no plans beyond that point. He didn't know where he was going to work or what he was going to do. God showed him it wasn't necessary for him to have a plan. God had this covered. In the year since his dad had the heart attack and God had shown Jeremy that He (not Jeremy) was in charge, everything fell into place. He was accepted into seminary, began the path to become a chaplain, and a job was provided to him. None of these things had Jeremy even considered.

God gave all of them to him. All he had to do was follow. *"**Trust in the LORD with all your heart and lean not on your own understanding.**"* Proverbs 3:5 (NIV) This verse continued to hold true.

But the good news didn't stop there. He drove back to Arkadelphia, and the next day, his parents came for a visit. Jeremy told his parents everything that had happened in the past few days, and then Jeremy got the best news of all. A pastor he knew was driving by his parents' house and felt the need to stop and visit Jeremy's parents. During the course of this visit, his dad was saved and became a child of Christ. Another of Jeremy's prayers was answered, showing Jeremy that God was in charge and things happened in God's time, not his. This was a lesson he has not forgotten. He started his new job at Camp Robinson, and his office had a window that allowed Jeremy to look at the chapel for the next three years while he finished up his education and the long process of becoming a chaplain.

Chapter 15
GO WEST

Two years into seminary training, Jeremy was again asked to take the position of youth pastor of a church that was, frankly, struggling. Jeremy accepted the position, and he and Beth began the work God had put before him. Of course at this time, they weren't alone. The couple had moved back to Greenbrier, began remodeling a house, and found out their first child was on the way. Jeremy worked tirelessly and began to grow this ministry as his family also began to grow.

The youth group outgrew their first classroom, then outgrew the fellowship hall and had to be moved into the sanctuary. Jeremy and Beth experienced successes and failures during this time, meaning they were not able to reach all of the kids. But there were success stories, e.g., one of the kids is now a chaplain's assistant in the Navy and another runs a small ministry beside her husband and kids. It buoyed them to see these successes and helped lessen the hurt of the failures.

Jeremy was invited one night by a friend of his in seminary to listen to a man preach during a revival. This exposed him to the Bible teachings of Dr. Travis Plumlee. Jeremy had never heard preaching and teaching like that. Without even knowing Jeremy, Dr. Plumlee told the pastor of the church he had found the youth pastor they were looking for and pointed at Jeremy. Again, this man knew nothing about Jeremy but discerned the plans God had for him. On the spot, the pastor offered Jeremy the position. He also asked Jeremy to become a BOB, which turned out to be one of the most important things in Jeremy's life to this day. This Band of Brothers was a continual, intense discipleship ministry directed at selected ministers.

There were five men in the initial group that Jeremy attended. He was introduced to the concept of keeping a daily journal. The first thing Travis told Jeremy to write in his journal was GO WEST. It stood for Get Out, Witness, Evangelize, Serve, and Testify. This was the basic premise of everything Jeremy was taught. Jeremy observed Travis speaking at different churches and discovered people either loved Travis or hated him. The reason they hated him was because he would call people out on their relationships with God. He reminded Jeremy of the Old Testament prophets who would stand on the temple steps and call out the religious leaders of the day. This group of BOBs learned as they watched him. He wasn't a celebrity pastor like those seen in many churches today. He was a Bible teacher who was humble until he saw sin, and then he would attack it with a ferocity and strength that was absolutely based on the Word of God.

Toward the end of this three-year period, the State Chaplain job for the Arkansas National Guard became available. Jeremy called Travis and sought guidance on what he should do. Travis told Jeremy this was what he had been called to do and to go forward. He told Jeremy to pray fervently and make sure this application was sent to the proper channels, and Jeremy followed this guidance. Keep in mind he hadn't finished seminary, even though he was close, and had never served in any sort of chaplain position at all. In fact, Jeremy's total ministry experience consisted of being a youth minister at a few small churches.

All of the other applicants had chaplain experience, some with experience through their entire military careers. Jeremy was a first lieutenant (O-2) and all of the other candidates were colonel (O-5) and above, some having been deployed with the soldiers. He was asked to interview with the

Adjutant General of the State, the Chief of Staff for the State, and the Arkansas National Guard State Chaplain. They asked him the question, "Why should we hire you?" and Jeremy began to explain the teachings to which he had been exposed in the past year. The adjutant general was the most aggressive (not in a bad way) interviewer and presented several scenarios, asking what Jeremy would do in each one. They asked him how he would do things differently than what had been done. Jeremy's response was, "I'm going get out and walk among the troops. I'm going visit them. This is a base of operations, but I don't have to stay within these four walls. The ministry can be so much bigger than it is now."

This reflects the first premise of the Great Commission (Matthew 28) where Jesus tells his disciples to go to the people and preach. Jeremy also let them know that his youth presented an opportunity. He could be molded, developed, and helped to transform the chaplaincy in Arkansas which at the moment had only four others serving in roles. He left this interview and proceeded with his daily life, doing his job and focused on completing seminary.

About four weeks later, he got a phone call from the Arkansas National Guard Human Resources office letting him know he had gotten the position. He remembers nearly dropping the phone in his lap upon receiving the news, and on April 17, 2007, he became the full time Support Chaplain for the Arkansas Army National Guard. At the same time, he had graduated from seminary and had a date to attend the Chaplain Basic Officer Leaders Course in Columbia, South Carolina.

Chapter 16

The Fun Begins

When Jeremy walked into his office on his first day on the job, he found three things sitting on his desk. First thing was a CD from his predecessor. The CD contained all of the files from the predecessor's computer. The second item was a packet of materials he needed to get together because he was short of staff. He needed to fill the civilian administrative (admin) assistant position, and the job packet had not even been put together. The office is a three-person office: the Support Chaplain, a Chaplain Assistant, and the civilian admin assistant. The Chaplain's Assistant when Jeremy took over was Sergeant (SSG) Eric Flud. SGT Flud was a civilian Pentecostal preacher and very much the polar opposite of his new chaplain. That proved to be a good thing. As they went through the first year, SGT Flud was a rock in the office and definitely kept things going as Chaplain Miller was learning on the job.

The third thing on the desk was the results of an investigation and audit of the Chaplain's Account for misappropriation of funds. It turned out there were twenty confirmed instances of misuse of the funds. It sounds much worse than it was, and in less than a year, Chaplain Miller had cleaned up all instances of misuse and put together procedures to ensure these did not happen under his watch.

On the first day, SGT Flud informed Chaplain Miller he (Miller) was preaching on Sunday morning. Jeremy spent the rest of the day organizing his office and praying in the chapel. The rest of the week was spent familiarizing himself with the surroundings and taking stock of what forces he had at his disposal. This was a great first step because very quickly those forces were going to be called upon. His first priority was to determine what was needed on Camp Robinson, the primary base for the Arkansas Army National Guard. Chaplain Miller had resigned all of his civilian ministry positions upon taking this position and was able to commit all of his attention to it. This also proved to be a good move on his part.

There were three programs involving at-risk teenagers on Camp Robinson at that time. One was a GED Plus program where students experience both a military basic training and a structured academic environment. Another program was Youth Challenge (YC). YC is a 22-week residential program for 16- to 18-year-old at-risk youth in desperate need of behavior modification. The program is supported by the military but run entirely by civilian teachers and cadre. The final program was the Civilian Student Training Program or C.S.T.P. This was a state-funded and court adjudicated program providing behavior modification instruction in a structured, military-style environment. It accepted 13- to 17-year-old males only. Each young man was a nonviolent offender in this nine-week residential program that stressed value-based learning, academic and life skills education, physical fitness, and community service. Chaplain Miller was to provide counseling and worship services for all who wished to attend.

The first Sunday morning, Chaplain Miller's expectation was to have a service for a few teenagers. His expectations quickly changed. That first Sunday, he ministered to over 250 at-risk teenagers. YC bussed their charges to the chapel and C.S.T.P. marched theirs. They sat on opposite sides of the chapel, one side wearing green shirts and the other side wearing all blue. Then, the female members of the YC program were brought in, and they were seated upstairs in the balcony. There was absolutely no mingling among the groups and between the boys and girls of YC. Chaplain Miller preached his first sermon, and it was a strong one. One of the cadre (individual in charge of the youths) asked him if every sermon was going to be like that, and when Chaplain Miller replied it was, the cadre stated he was glad to hear it because it was what those teenagers needed to hear. Jeremy preached the love, grace, and mercy of Jesus, and for many of the teens, it was the first time they had heard this message. He offered Bibles to those who wanted one and he exhausted the chapel's supplies on the first Sunday.

As those groups were leaving, a bus load of soldiers arrived and asked if he could preach a service for the GED Plus Soldiers. Chaplain Miller preached the same sermon to approximately twenty-five young people. At the end of that service, one of the cadre informed the Chaplain that by the end of the summer, they would be receiving approximately five hundred people every three weeks. In short, the program was about to explode. They requested chaplain support and asked if he would be able to handle it. Eventually he helped them get their own specific chaplain, but for years until that happened, Chaplain Miller was their resource for support.

Chapter 17

Honor The Dead

There are three pillars that are absolutely necessary for Army Chaplains: 1) Nurture the living, 2) Comfort the sick and wounded, and 3) Honor the dead. This third pillar was about to become reality.

At the same time Jeremy assumed the Support Chaplain position, the 142nd Field Artillery and the 875th Engineering Battalion of the Arkansas National Guard were being mobilized to Iraq. Monday morning, after preaching to the young people the day before, Chaplain Miller received his first phone call from Mortuary Affairs, Captain Marty Patterson, notifying him there had been a casualty among one of the units deployed to Iraq. Chaplain Miller was informed he had to do a death notification. To say he was totally unprepared for this would be a vast understatement. He had never done one of these, practiced for one of these, or studied in seminary how to do a death notification. But it was part of the job and he would perform his duties.

Chaplain Miller got a little bit of a break because the brother of the deceased soldier was also in-country and had notified the family prior to the death notification team arriving. The deceased soldier had requested to have his sister notified first and then his mother and father. When the team arrived at the apartment of the sister, her husband greeted them at the door and informed them they had been expected. They were able to perform their duties, offer comfort and information to the family, and try to answer as many questions as possible. The one thing that really stood out that day was as they were leaving, the sister hugged the chaplain's neck and told him thank you. This resonated to him in a way he didn't understand at the time. He thought, how can they thank me when I've just delivered the worst news they possibly could hear? That question got a lot deeper very quickly.

About three days later, Chaplain Miller got the second of these phone calls. He and the Mortuary Affairs Officer drove to Searcy, Arkansas, to notify the wife and children of this deceased soldier. One thing people don't really understand is the type of home environments a large number of these soldiers come from. Television has given us a distorted picture of military death notifications. We watch movies and television shows showing notification teams arriving at nice homes and large farmhouses to notify parents and wives. This is so not the picture, especially with guardsmen. Most of them have joined the Guard to supplement their civilian incomes, and a lot of them do not live in upscale neighborhoods or farms. These soldiers were not born with silver spoons in their mouths and such was the case in this instance.

The team drove into a trailer park, and not one of those well-kept trailer parks. They parked a little away from the deceased soldier's trailer and walked through the park towards his home. As they walked down what best can be described as a gravel street, people began to come out of their trailers and watch as these Army soldiers in dress uniforms walked by their homes. No one was dressed up, so the deep blue military dress uniform with its shined buttons, medals on the chest, and patent leather shoes made for an odd sight. They got shouts of, "Where you going?" and, "Who you looking for?" from the people, but they didn't answer and just kept walking.

They arrived at the family's front door and knocked. The wife, a day earlier, had been told her husband had been wounded, and this was the last report she had received. She opened the door and the officer with Chaplain Miller began to tell her, "The Secretary of the Army regrets to inform you of the passing of your husband," and that's as far as he got. She fainted dead away, and Chaplain Miller caught her before she hit the ground and was able to place her in a chair sitting on the rickety deck by the front door, immediately praying for her in his mind. It took a little bit for her to come around and they noticed people starting to gather, watching this scene unfold. To Jeremy, they watched not out of concern but possibly out of entertainment to fuel gossip later on.

Chaplain Miller stood and for the first time noticed all of her children standing at the door. There were several children, none of them very old and most of them with no shoes and little dirty faces, with at least one in diapers.

As the wife came to, she was emotionally distraught, and Chaplain Miller asked the officer to kneel down next to her and tell her what he knew. He and the mortuary officer were both trying to keep it together, but that didn't last long. As he stood and moved back from the spouse, he felt a tug on his jacket sleeve. He turned around and this little boy, about six, blond hair, blue eyes, dirty face, and no shoes was looking him straight in the eye. The little boy, holding on to Chaplain Miller's pants leg now, looked at him and asked, "Do I have a daddy anymore?" Chaplain Miller's eyes filled with tears, and he replied, "No, buddy, I'm so sorry." With those words out of his mouth, the surge of emotions hit him. The feeling of that moment haunts Chaplain Miller to this day. Of that time he says, "I was done. I was crying and sat down on the deck. We all were."

The little boy sat down next to him on a little stump that was on the deck and looked at him. Then the little boy asked, "Is my daddy in heaven now?" and Chaplain Miller simply didn't know what to say because he didn't know the soldier and didn't want to lie to the little boy. Chaplain Miller said, "Buddy, there's a way to know if we are going to heaven and that's to trust in Jesus," and prayed with him right there.

They stayed with the wife and family for a while and when they felt they could leave her, they headed back to the truck. The crowd in the park was getting larger and all eyes were on them as they walked. They still had to go inform the mother and then the father of the soldier. They asked the wife not to call the mother for a little while until they had made the notification.

One soldier, but three separate, emotionally draining notifications. It was a rough day, but later in the evening, the wife called Chaplain Miller and asked him to do the funeral for her husband. She informed the chaplain her husband had become a believer about six months before he was deployed and was a member of a local church but thought it would be good for the chaplain to do the service. She wanted Jeremy to handle things because she saw the care he had shown her and her children earlier in the day. She had also spoken to her in-laws and had the same report. She knew he would bring honor to her husband.

It was a service with a strong military contingent and a military chaplain leading it. The soldier had asked that "Free Bird" be played during the viewing at the end of the service, and Chaplain Miller thought it was extremely fitting. Southern boys and Lynyrd Skynyrd. Not much else needs to be said.

The first four months Chaplain Miller was on the job, he performed this duty fourteen more times. Not all were from the Arkansas National Guard, but as he was the primary chaplain for the Army in the state, he was asked to do this notification for other branches, as well. The signature thing that marked Chaplain Miller reflecting the third pillar of all chaplains, honoring the dead, was the fact that he never hesitated to perform this duty. Not all were pleasant. He was cussed, degraded, and in one instance, assaulted. But in every instance, he continued to love the family and try to show the mercy and grace of Jesus.

Chapter 18

Slowing Down

After the fourteenth notification in four months, Chaplain Miller had emotionally had enough. His emotional state was a mess and he needed a break from this duty. Every notification wasn't just for one person but was performed for several members of the family depending on their dynamics. He was having to be the rock for the mortuary officers doing the actual notifications. He was driving the vehicle to and from these notifications and in some cases that meant driving to neighboring states. He was tired, both mentally and emotionally. He went to his upper leadership and told them he needed a break. But true to the military credo, his commander told him to suck it up. He was all they had at the time and he was it, get it together, and keep doing the job he was doing.

There is a term prevalent in the military: embracing the suck. What this means basically is "yes, it's hard but get over it." A better definition is the best way to get through these times or experiences is to embrace them instead of ignoring or denying them. This was Chaplain Miller's first true experience embracing the emotional suck, but would not be his last.

Jeremy realized he was looking in the wrong place. Again, he asked God to help him and turned it over to God to deal with, and deal with it He did. Chaplain Miller learned how to rest on the run but better than that, he was able to hire his first civilian admin assistant, Ms. Deborah Beal.

One day when Chaplain Miller arrived for work, SGT Flud, his military chaplain's assistant, met him and asked the chaplain to give him his cell phone. Flud was going to put it on charge for him and would bring it to him when fully charged and ready. Chaplain Miller sat down at his desk and began to take care of the day-to-day tasks that have a tendency to get pushed to the side. It got to be lunch, and Chaplain Miller suddenly realized he hadn't talked to anybody and wondered if his phone was charged. When he asked Flud about it, the sergeant politely informed him the phone was fine and no messages had come across it he or Ms. Deborah couldn't handle. In addition, he had put a sign on the door stating the chaplain was out and put his own phone number to be contacted. He literally forced Chaplain Miller to take a break. When asked why, SGT Flud responded, "You are my responsibility. If you go down, we all go down. I'm your armor bearer and it is my job to take care of you. So there will be days when I do this and I will make sure you get these types of days when you need them." Frankly, it is what a good NCO does for his officer, and SGT Flud was a good NCO. On those days, Flud or Ms. Deborah took all of the phone calls and talked to all of the people. They would notify offices not to bring people to the chaplain's office and would force rest days on Chaplain Miller. This was answered prayer and the result of letting God take care of things instead taking care of them himself.

Because of this action and having a fully staffed office, Chaplain Miller was able to begin to get things truly organized. They were able to develop a system to ascertain which items were truly important and which items were not, and they use that system to this day. It has enabled the entire staff to be able to focus on the items and issues they need to focus on and eliminate those they do not. At the end of 2007, the three items on Chaplain Miller's desk were completed. He didn't need the CD, he got his chaplain admin assistant hired, and the audit had been cleared up.

It was a hard but worthwhile accomplishment by the six month mark. Looking back on it, Chaplain Miller was amazed by where they had been and where they had gotten to. His own office was now fully manned and continues to be. He has grown the chaplain ranks for the Arkansas Army National Guard. When he became the support chaplain for the state, there were only four chaplains in the entire state guard, counting himself. Years down the road, they would have nineteen chaplains in various positions throughout the state units. He has made it a priority to fill every chaplain position available to the state and then some. This ensures the type of chaplain coverage every soldier needs and deserves. In addition, since there are almost continuous deployments, it enables chaplains to be deployed more often with the troops, something that is desperately needed.

Chapter 19

Another Deployment Missed

In 2008, the 39th Infantry Brigade was mobilized a second time to Iraq, and for the second time, Chaplain Miller was told by the powers that be that he was not going with them and there was nothing he could do about it. He was devastated. These were his troops, the ones he ministered to on a daily basis, and when they left, they were leaving their chaplain behind. But it was a bad news/good news scenario because even though he was not going with the troops, the ministry on the post was beginning to explode in one significant way. There was a marked increase in the worship services through 2008 and leading into 2009. Because of the other programs previously discussed, the number of worship services moved from one a week to five worship services each Sunday, plus Wednesday evening services, and the numerous Bible studies being conducted. Because of the uptick on the number of services, Chaplain Miller was able to expand his staff by one more chaplain. To keep from burning out one chaplain, he was able to get two new chaplains to the state and put them on alternating orders. Bill Berry and Carter Pierce, good friends of Chaplain Miller's and guys he said could "preach circles around me," stepped up. They proved to be absolutely awesome chaplains and helped Chaplain Miller in so many ways. Chaplain Pierce has now retired from the Guard and Chaplain Berry, after a stint in Kentucky, has come back to Arkansas and is again a member of the Arkansas National Guard.

Having five services on a Sunday was not an easy undertaking. Sometimes one chaplain would preach all of the services, and sometimes they would split them up. The day would start with an 8am service with the first round of GED Plus program individuals. That service would hold between 200 and 225 individuals. As that service ended, the second round of GED Plus folks would be arriving for the 9am service. This was a smaller batch, usually between 75 and 100 troops. The 10am service was for the Youth Challenge/CSTEP students. The Youth Challenge/CSTEP services were held at 10, 11, and 12 noon. The young people in these programs began making professions of faith and trusting Jesus Christ as Lord of their lives and asking to be baptized.

This presented a couple of challenges for Chaplains Miller, Berry, and Pierce. Good challenges, but challenges nonetheless. First, when these young people finished these programs, one of two things happened: 1) they joined the National Guard and prepared to ship off to boot camp or 2) they went back to their home towns. For those going back to their home towns who had made professions of faith, the chaplains worked to get them plugged into a church upon their return. This involved a lot of phone calls and additional time being spent getting these young people into a church. Again a good challenge, but still a challenge.

They would ask the young people if this was the first time they had been in church and over 80% would raise their hands. Same was true of having read or owned a Bible. This was a truly unchurched group of individuals and a ripe mission field. Those who professed a faith in Jesus and wanted to be baptized got baptized. There is a lake on Camp Robinson, and they used the lake to conduct baptism services. There was no set time or day, but when it could be fit into the schedule, baptisms happened. Chaplain Berry assumed this responsibility, and GED Plus eventually hired him as their full-time chaplain. Jeremy would finish his services and drive past Hunter Lake to watch the "Bill the Baptizer" with students lined up. When Chaplain Miller looks back on this time he now understands why God chose to keep him home and did not allow him to deploy at that time. There was a field ripe for God's Word and ministry, and God used the three of them to work that field and bring young people to Christ who otherwise may never have had that opportunity.

One individual in particular sticks out in Chaplain Miller's mind. Jeffrey Jordon was one of those individuals who accepted Christ in one of the services and he later joined the National Guard and deployed to Afghanistan. He was killed in action during that deployment and now has a training center on Camp Robinson named in his honor. Soldiers step into harm's way for their country every day. Some pay the ultimate sacrifice. The work of these chaplains help soldiers know they can spiritually find peace in their missions if they are called upon to pay that heavy price.

Chapter 20
Suicide

It was not all good experiences when staying behind from a deployment. There was also a very bad part Chaplain Miller and his staff would have to deal with. When the 39th returned in 2004 and three years later began preparing to deploy again in 2009, a nationwide problem hit the military. Suicide rates of soldiers previously deployed to Iraq began to rise, and the Arkansas National Guard was not exempt from this increase.

Suicidal ideation is thinking about or an unusual occupation with suicide. Chaplain Miller and his staff found themselves spending more and more time with soldiers and veterans (discharged soldiers) coming to them seeking help. Not only were soldiers and veterans thinking about suicide, the number of attempts rose dramatically. The adjutant general for Arkansas at that time recognized this problem as well, and he organized a task force within the Guard to deal with this issue. Unfortunately, this task force was not what you would picture as a task force. The initial team was Chaplain Miller and a female captain from family services.

Because it was just the two of them did not mean they didn't work hard on this issue. Both put out 24-hour contact numbers and began attending every class and seminar they could find on the subject to try to educate themselves as much as they could to deal with this issue. Frankly, this is not something any ministerial education system spent a great deal of time on and this included all of military chaplain education. So, they had to get this from other sources, and they did. Once the 24-hour contact number was made available, they began to get phone calls at night, every night.

One call in particular sticks out to Chaplain Miller. Beth was pregnant with child number three and the phone rang after they went to bed. There was a sergeant on the phone and he wanted to talk. The first words out of his mouth went something like this: "Chaplain, I've taken a ton of pills. I'm starting to get drowsy, and I want to give you my last will and testament over the phone. I want you to pray with me before I die." Chaplain Miller tried to get the soldier to tell him where he was but the soldier was reluctant. He did say he had prepared an email to his friends and family that was scheduled to be sent after he died. Chaplain Miller grabbed his other cell phone and started texting the captain and letting her know what was going on.

The conversation started to stretch out and after about an hour, the soldier was becoming very groggy. Chaplain Miller continued to pray with and for him and also began work on getting his location from him. Beth was still sitting in the bed, praying for this soldier. She was worried, but this event had been playing itself out time and time again in the recent days. She knew Jeremy was trying his best with limited resources. Because he was able to keep the caller on the phone, the captain was able to get his cell phone pinged by local police. Suddenly the soldier said, "I've got to lay down now, thanks, Chaplain," and he set the phone down and ceased communicating with Chaplain Miller. Jeremy was screaming into the phone to get him to sit up and speak. Then he heard the best sound he's ever heard. Through the phone, Jeremy heard the sound of the soldier's door being kicked in and paramedics arriving on the scene. They got to him and got him to the hospital in time to save his life. This was one of the success stories. Sadly, way too many do not end this way.

Chaplain Miller recalls cases where they never got the call. He remembers nights when a fellow chaplain slept outside the door of a soldier's room who threatened to kill himself. He also recalls trips to the Veterans Hospital with men and women seeking help. He would go to visit them only to find out they checked out, stopped by the bar, and loaded up with enough liquid courage to take their lives. It is devastating to lose soldiers, and each one of those losses is haunting.

Through the efforts of this team and the growing task force, they were able to train others and establish a network of outside agencies to assist them with this issue. In fact, the captain has since discharged from the Guard and now works with the state of Arkansas on suicide prevention and continues to be an asset for Chaplain Miller. Between the two of them, they traveled the state educating everybody they could about this issue exploding within the military community. Again, with all of the successes in this area, there is one failure that continues to work against them.

That first year, in three separate incidents, they referred soldiers to the VA hospital in Little Rock because these soldiers were threatening suicide. They asked that the soldiers be put on a 72-hour hold, which was granted. During that three days, chaplains and others would visit with the soldiers, as would the counselors at the VA. At the end of the 72 hours, the VA didn't even have the common decency to notify anyone the soldiers had been released. They didn't notify the families, the Guard, or the chaplains. They just released them at 5pm, happy hour. In each instance, the soldier would go to a bar, have a drink, and go home and take their life. There was nothing the task force, families, or other soldiers could do. When confronted with this, the VA's response was you can't tell us we have to tell you anything.

This was elevated to the top of the VA system in Little Rock, and in Chaplain Miller's words, "to work through that system is insane."

The Guard leadership in Arkansas fully supported anything that would lead to fixing this issue. They blessed off on a plan that had Chaplain Miller's team and the head of the VA in Little Rock come to a written agreement that basically said any time a soldier is brought and/or referred to the VA hospital because of suicide ideation, before you release them, you will notify the Guard representative. The Guard wants to be involved in every level of treatment and want to know before they are released so we don't continue to lose soldiers in this manner. Simply putting this program into place between the VA and Arkansas National Guard helped bring the suicide rate of deployed veterans to zero by 2010. But it took having to hammer on the head of the VA in Arkansas to accomplish this. That number has begun to creep up again in recent years, but things are being handled now in a much better manner.

Chapter 21

Westboro and Lobster

In the 2009 deployment, there were two casualties and neither was combat related. One was a heart attack and one was the result of a traffic accident. The funeral for the soldier who died as a result of the traffic accident was one that Chaplain Miller says he'll never forget. Chaplain Miller has had the unfortunate pleasure of having two of his military funerals picketed by a group from the Westboro Baptist Church. [1]

This soldier was from the northwest Arkansas area, the funeral was conducted in Fayetteville, and he was laid to rest in the Fayetteville National Cemetery. The family had received warning the picketers were going to be there. The picketers had obtained the necessary permits and they were stationed across the street. They also wanted to know the route the funeral home was going to take to the cemetery so they could go there as well. Legally, that is their right, and soldiers fight so they can have that right. They lined the street across from the church with signs calling troops evil or holding up even more derogatory signs. You could hear them

---

[1] AUTHOR'S NOTE: Frankly, I am ashamed that people claiming to be Christians act this way. In my opinion, they are the Christian equivalent of the Taliban and definitely do not understand the teachings of Christ. They meet every definition of false prophets. I have some other opinions about them, but let's just say I would lose no sleep watching each and every one of them be locked up. Having had a son deployed in both of the Arkansas 39th Infantry deployments, and my own retirement from the Arkansas Air National Guard, I find their actions disgusting. END AUTHOR'S NOTE.[1]

screaming that “God loves dead troops” or random out of context verses from Revelation.

Chaplain Miller stared them down as he waited for the family. He imagined what the headlines would read if he crossed the road and started breaking their signs and flipping their tables. Thankfully, his focus shifted with the arrival of The Patriot Guard Riders (volunteers made up of mostly former military as well as civilians who provide motorcycle escort). They met Chaplain Miller and spread out to form a line of huge American flags that prevented the family from seeing the picketers. They provided a phenomenal service to this family and continue to do so for other families of fallen soldiers.

These men and women were mostly from the Vietnam era and had felt the full brunt of a nation who distained their military service. Their job, their ministry was to stand in the gap and save this family any sense of shame during this time of loss. The head of the police detail came to Chaplain Miller and informed him he needed to ride in the lead vehicle with the family once the service ended due to having to drive past the picketers. Chaplain Miller counseled the family of the soldier about what might happen and advised them to continue to make this day about their family member and not let idiots ruin this day.

The funeral went off without a hitch, and when they came out of the church, Chaplain Miller noticed the picketers were gone. He asked about it and was told the picketers had asked for a police escort to the cemetery since they didn’t know where it was. Everybody left the church, and when they got to the cemetery, there wasn’t a picketer in sight. The graveside service went off with no problems and everybody left. Chaplain Miller rode back with the family and they dropped him off at the church to pick up his vehicle. He talked to one of the policemen still at the church about what happened and the officer smiled really big. He said, "Chaplain, we have a new deputy on the force and he gave the picketers their escort. He's not real familiar with the area and accidentally drove an hour in the wrong direction. By the time he realized his mistake and was able to get them back, the funeral was over." Again, all of this was said with a big smile. Chaplain Miller cannot talk about this without laughing and being appreciative for whomever that deputy might have been.

There are also funerals that do not go as expected. One day, Chaplain Miller received a call from a friend of his in the Guard. This soldier told him he had a relative pass away and needed some help with the funeral since the deceased was a veteran. He trusted Chaplain Miller and asked if he would mind conducting the service. The relative wasn't religious in any way but was a veteran and that was all that mattered to him or to any other chaplain. The family was scattered and the girlfriend was a mess, drinking herself into a stupor just before the service began. Chaplain Miller worked with the soldier and they got the service arranged with Chaplain Miller doing the flag presentation to the girlfriend, which is not normal, but also not uncommon. Just before the service started, the soldier came up to Chaplain Miller and advised him the girlfriend was a little hysterical and on the tipsy side (understatement). Chaplain Miller told him he had dealt with these types of situations before and the soldier looked him in the eye and said, "I don't know if you are prepared for this." Chaplain Miller got a sense the soldier was trying to warn him something was about to happen, but again, Chaplain Miller advised him he had everything under control and nothing would surprise him. Then he got the opportunity to eat those words.

When the family walked out, the girlfriend, under heavy support, walked with them and in her hands she was carrying a singing lobster, the kind you see on a wooden base where you push the button and it plays a tune and the lobster moves - the claws shimmy and shake. This was definitely out of place at a funeral. The girlfriend walked past Chaplain Miller who was standing in front of the flag draped coffin and set the lobster on top of the coffin and returned to her seat.

Chaplain Miller left it alone and proceeded with the ceremony. He went through the entire service with the eulogy, the sermon, and the military honors team, and throughout, the girlfriend was visibly upset. The military honors team gently removed the lobster, finished with the flag and gave it to Chaplain Miller to present to the girlfriend. He took the flag and knelt before her saying, "This flag is presented on behalf of a grateful nation for the services and deeds of your ..." and as he said "your" the girlfriend grabbed him by the arm and pulled him in real close. As she did this, Chaplain Miller could smell the strong odor of alcohol on her breath. Her speech was heavily slurred but she then said, "Say 'your lobster'." He said, "Excuse me?" and she repeated, "Say 'your lobster'. He was my lobster. I don't want to hear his name. I want you to say lobster!" She released his arm and Chaplain Miller reassumed his posture and began again. "Ma'am, this flag is presented on behalf of a grateful nation for the services and deeds of your ... lobster." Chaplain Miller looked over the girlfriend's shoulder and made eye contact with the soldier. The soldier put his head down and was shaking it as if to say, "I tried to warn you."

Chaplain Miller stood up and started to salute the flag. The girlfriend immediately shot out of her seat, ran past him, and before the honors team could exit, got to the lobster and pushed the button to activate it. The lobster started to dance and she yelled, "Everybody dance with me. This is for my lobster." Chaplain Miller dropped his salute, did an about face, grabbed his Bible and just stood there as this woman danced with the singing lobster. Again he looked at the soldier and now other family members who were mouthing, "I am so sorry." The family was mortified. The honors team and Chaplain Miller exited the building, and the funeral broke up at that point. Everyone mourns in their own way. Jeremy was upset or even embarrassed. He came there to share God's word of hope with the family and to honor a veteran for his service. Even if rendering that honor followed his salute with an under the sea dance party for one.

Chapter 22

New Chaplain

In 2010, Chaplain Miller became the brigade chaplain for the 39th Infantry Brigade. He discovered he was the only chaplain in the entire brigade. There are four brigades within the Arkansas National Guard, and each brigade has a lead chaplain. Within each brigade there are battalions, and each battalion also has a chaplain assigned to it. So when Chaplain Miller took over the 39th, there were no battalion chaplains underneath him. Chaplain Miller, because of his role as the support chaplain for the Arkansas National Guard (which is a civil service position), now additionally found himself in the role of having to recruit chaplains for not only the battalions of the 39th but also battalion chaplains for the other three brigades comprising the Arkansas National Guard. The primary reason for the shortage of chaplains in the brigades was they had gone through two eighteen-month deployments in less than five years, and chaplains were leaving the Guard upon completion of these deployments. Being short of chaplains was an issue because after the brigade returned in 2009, it began to encounter a myriad of issues. Troop strength declined, the suicide issue began to manifest itself, and PT standards weren't being met. Two full combat deployments in a five-year period took its toll. He knew he wasn't alone in seeing the issues. Many wonderful officers and NCOs across the state were working hard to "right the chip," if you will.

During this time, Chaplain Miller recruited, in his words, "some of the most incredible chaplains I've ever known". Of the chaplains about to come on board, three later became brigade chaplains and two others were selected to become active duty chaplains. Getting the chance to mentor these chaplains, teach them the ropes and mold them into military chaplains has been one of the most satisfying experiences of his life. One of the things he always told them was: "Whether you are standing on a mountaintop, waist deep in mud, or counseling a soldier because his wife left him because of his third deployment, do your job as a chaplain and care for your troops." His chaplain numbers began to rise and, as noted earlier, some very strong chaplains become part of the brigade. He also was able to recruit some strong chaplain assistants, which is germane to each chaplain and his effectiveness.

Upon arrival as the brigade chaplain, Chaplain Miller found himself outranked by soldiers in the other staff positions. He was a new captain and everybody in an equivalent staff position was a major or above. He made his mark on the staff with one of his strengths. The brigade commander loved PT (physical training) and this hit Chaplain Miller in his wheelhouse. He came to Chaplain Miller and asked that he do services that everybody could attend. So Chaplain Miller used PT as a ministry outlet. On drill weekends, the brigade staff would meet at Pinnacle Mountain (one of the small mountains in Central Arkansas) and as a staff, they would run the mountain. They would begin in the dark (5 – 5:15am) and as they ran and climbed this mountain, the sun would begin to rise. When they reached the summit, the sun would be up. Once they reached the top, Chaplain Miller would do a devotional with the staff. That became synonymous with the way they began to do worship. If you wanted to do worship within the brigade, you had to climb a mountain. He began to meet non-believers and weak Christians this way because he was on that mountain and on that trail running and sweating with everybody else. The fun part of this became Chaplain

Miller trying to get back to the camp on Sunday morning to preach the eight o'clock service because the senior chaplain for the state on the military side would go back to his home church to preach.

Chapter 23

NTC

In 2011, the brigade again found itself on the verge of deployment, this time to Afghanistan. To prepare for this deployment, the entire brigade was sent to the National Training Center (NTC) in Fort Irwin, California, for summer camp. NTC is an enduring installation recognized for providing a realistic joint training environment focused on honing warfighting skills in hostile environments, i.e., Afghanistan. They do an excellent job of this. In fact, armies from around the world come to NTC to sharpen their skills against a resident opposing force that knows the terrain and every trick in the book. Once a unit has completed their training at NTC, they are prepared to face whatever an enemy can bring to them. NTC has adapted their tactics to be able to simulate whatever enemy a unit might be facing when deployed.

Before a unit can deploy to NTC, the leadership of the unit must attend and pass a Leadership Training Course (LTC). All of the commanders and staff of the brigade and accompanying battalions have to attend this course and they war game every scenario for a week. A chaplain has two roles during this training. There is a real-world role and the fictitious role-playing role. The chaplain must be able to advise a commander in all aspects of an operation in which religion might play a role. This is especially important considering the role religion plays in the Middle East, which at that time was the focus of most U.S. military operations. A chaplain not only needs to know his own religion but also must be familiar with the prevailing religion of the deployed region. It is called a Religious Area Analysis. Having this knowledge also helps the lead chaplain know how to equip his chaplain staff, not only in knowledge but also in material and supplies.

For example, some followers of Islam bundle up a bunch of switches and beat themselves on the backs and then begin to walk barefoot. This is a form of religious devotion to tell their god they are sorry for their failures and willing to punish themselves to find favor with the god. They would be bleeding and bruised. The chaplain would advise the commander that if troops went into that area, they would be encountering individuals who were bloody and looked worse for the wear. Letting the troops going into the area know they would be encountering these individuals could do a couple of things. One, they could let the imams – the Muslim religious leaders - in the area know there is an aid station for people who might need it, and two, these people probably don't need to stopped and interrogated just because they are bleeding. This is an example of a situation a commander might encounter and of which he needs to be aware. With that advice and knowledge, proper decisions can be made.

One of the scenarios at NTC involved the leadership being sent to meet with a number of Afghanistan leaders. (NTC hires immigrants from the appropriate countries to play these roles.) There was no one from the leadership group below the rank of lieutenant colonel except Chaplain Miller, who was a captain. The group of Afghans met and spoke with the leadership and listened for a while. When it got quiet, one of the mullahs – a Muslim mosque leader - stated they would like to hear from the military's imam, which was Chaplain Miller. He suddenly realized that everybody in the room was looking at him and he proceeded with a cordial conversation. At the end, the Afghans told all of these officers they should let their imam speak more often. Again, this was the point of going through the scenarios. Captains don't usually have a lot to say in a room full of colonels without being asked to speak, and without that knowledge beforehand, Chaplain Miller might never have gotten the opportunity to participate.

The real-world role involves taking care of the troops when they've worked these exercises for something like 30 hours straight and find themselves operating on fumes. The chaplain watches these individuals to make sure they are not only in a position to do their jobs but also in a position to keep up with situations as they develop. For example, one individual's wedding anniversary occurred during this time and Chaplain Miller knew about it. He asked if the individual had called his wife about their anniversary and the guy admitted he hadn't even thought of it. Jeremy gave him his cell phone and told him to step outside and make that call. A little thing, but something that keeps a troop grounded and goes a long way at home.

Less than six months of annual training was conducted at NTC, making theirs the first brigade combat team from the National Guard comprised fully of National Guard soldiers to complete a rotation. Needless to say, there was a lot of pressure on everybody. Chaplain Miller was met by the NTC chaplain observer/trainer whom he described as being about 6'8" and towered over everybody. The observer/trainer team consisted of two other chaplains whose job was to rotate around and watch how the brigade chaplains performed their duties during the rotation. The lead observer was with Chaplain Miller the entire time. He would rate Chaplain Miller and his staff at the end of the rotation. Chaplain Miller's staff consisted of himself as the brigade chaplain and five battalion chaplains because they were finally at full strength. Chaplain Miller had a meeting with all of the chaplains and their assistants and informed them he might not see them again until the rotation was completed. He provided phone numbers, call signs, and email addresses on how to reach him. Chaplain Miller's call sign was Bowie Shepherd. The reason it was Bowie is because the patch for the 39th is a Bowie knife, the famous knife that was made in Arkansas for James Bowie. The second part of the call sign, Shepherd, is pretty much self-explanatory. He prayed with them and sent them on their way to their appropriate battalions.

Their initial accommodations were basic military accommodations. They all slept together in big tents or buildings and the showers were large, open showers. To go to the toilet, one had to go outside to a line of toilets that were facing the desert and where the wind never stops blowing. For privacy, one would hope to get a toilet with a shower curtain. If they were lucky enough to have a curtain, they would try to stretch it on the plywood walls to give some sense of privacy for themselves. Chaplain Miller described it like this: "You hold the curtain out tight with both hands. In the meantime, the wind is blowing it up against you and blowing sand underneath the curtain into your pants. Then this stark realization hits you that you have to hold this curtain, and when finished, you have to wipe. How are you going do this? How willing are you to let go of this curtain to stay sanitary?"

The unit stayed in these accommodations until they rolled into what is known as "the box." The box is terminology for combat zone, and this is where the real action happens. SGT Lang, Chaplain Miller's chaplain assistant while drilling with the 39th, was his driver, and along with them were two members of the Judge Advocate General (JAG) office. Together, they were called the Law and Grace team. The drive to the Forward Operating Base (FOB) actually took the convoy over four hours with no stops instead of the projected hour and a half. The FOB consisted of a walled compound, huge tents to sleep in (imagine circus tents), and military style tents for various functions. Other than the walls around the compound, there were no permanent structures. It was hot and the sand was blowing, meaning everybody had their goggles on.

Chaplain Miller and SGT Lang set up their small chapel and got their stuff organized. Chaplain Miller contacted all of the chaplains to make sure they got to where they needed to be, and everyone went to work. Again, real world and role-playing roles kicked into gear. Taking care of the troops as well as participating in the war games became second nature to all of the chaplains. One of the success stories during the deployment involved Chaplain Eaton, who was assigned to one of the battalions. Their unit was on top of a mountain and the Opposition Force (OPFOR) was going to try to take the mountain top. Chaplain Eaton was a firefighter in real life. He suggested that each of the squads carry a strobe light, and if they see something, they can activate the strobe and all weapons can be trained and fired in that direction. This was something nobody thought of and they were able to hold the top of the mountain using this strategy.

There were real-world things that happened as well because life continued at home. There were childbirths, parents passing away, birthdays and anniversaries. And with each of these, the chaplains were there to celebrate, cry with, comfort, or just listen to troops as those back home went through these experiences. It was, on a much smaller scale, exactly what happens when troops are deployed around the world. Try to imagine doing this while the entire time there is a 6’8” mountain of a guy breathing down your neck to make sure you do everything right. Operational briefings were able to show Chaplain Miller where his chaplains were located and how they were doing. He was not disappointed in their performance.

The team put in long days and nights. They battled stress not only from the operational activity on the battlefield but also from mental fatigue. The soldiers learned to push their bodies and the staff pushed their minds. Together, they found success.

Several weeks later, NTC was over and it was a job well done for the 39th. On the flight back, Chaplain Miller was sitting near the back of the plane. The command sergeant major (CSM) came back and asked that the Jeremy come with him. He followed the CSM and was led to the front of the plane. The CSM said, "The commander sent me to get you and bring you up here. He wants you to sit in a first class seat for the flight home because of the great job you and your guys did." As thankful as he was, it wouldn't have mattered to Chaplain Miller where he sat on the plane because he slept the entire trip.

The brigade continued to ready for deployment to Afghanistan. However, about six weeks after returning from NTC, the brigade got word their deployment had been canceled. It was a let down to the entire brigade after their great work at NTC, but such is the life of an Army soldier.

However, things were about to change, not for the brigade but for Chaplain Miller. As usually happens after any sort of deployment, pregnancies among the spouses of the troops, and the female troops themselves, began to rise, and the chaplain's office was not exempt. Two of the chaplain spouses become pregnant, including Beth who was now with child number four. By the way, this is normal for any military unit, active duty, guard, and reserve, after being away from home for any length of time.

Chapter 24

Why We Serve

Several weeks after finding out about Beth's pregnancy, the chaplain's office was notified that the 39th Brigade Special Troops Battalion (BSTB) of the Arkansas National Guard would be deployed to Afghanistan and it would require a chaplain. The BSTB is a battalion comprised of military police, military intelligence, engineers, and troops with specialized skills. Their mission was to support other units on the ground and to work closely with those units as needed.

The other chaplain whose wife was pregnant received the tasking for this deployment. However, there were issues with her pregnancy and it became readily apparent he would not be able to leave her for the time needed. One evening, the commander for the unit called Jeremy personally. He had spoken with the brigade commander and was requesting Jeremy personally for the mission. Even though his wife was also pregnant, Chaplain Miller volunteered to take his battalion chaplain's place. The decision was not made hastily, but involved much prayer and discussion between Beth and Jeremy. When they made the decision, they both felt that God was leading them in that direction and both acknowledged this was the reason God had taken Jeremy down the path he was on. All of the ways made easy and roadblocks removed to allow him to become a chaplain and to have the ministry he was leading became readily apparent. Chaplain Miller said, "This is why we serve. It's why all of us serve. When the time to mobilize is presented, it is what we do and do so readily. Yes, I knew my wife was pregnant, but I also knew in my heart that God was placing this in front of me because He had led me to this point."

The BSTB was headquartered at the Shelby Ross Armory in Conway, Arkansas. This became Chaplain Miller's second home as the unit prepped to meet the short timeline. On his first trip to the armory, Sergeant First Class (SFC) John Jeu, the Operations NCO, met Chaplain Miller, showed him around, and introduced him to the individuals he did not know. This was the first time he met his new chaplain's assistant (CA), SGT Chase Berecin. Chaplain Miller described Chase this way: "He was of Italian descent, short guy with slicked-back, black hair, and tattoos down both arms." Chase had never been a CA and had volunteered for the position. When Chaplain Miller asked about his experience, he replied he was a generator mechanic but could work on anything. He said he was resourceful and his scores on things that mattered (shooting and PT) were top notch. Shooting was very important to CAs because the chaplains do not carry a weapon and the CA is their means of defense. He saw this position as an opportunity for promotion and advancement. When asked what he wanted to do with the rest of his career, he had it mapped out all the way to becoming a sergeant major. When Chaplain Miller asked about his wife's views on this deployment, he replied she was on board and supportive. This was not his first deployment and that experience would prove invaluable. Everybody Chaplain Miller talked to about Chase gave him glowing recommendations, and Chaplain Miller knew God had given him the right man to work with him for this deployment. As he would find out later in the deployment, Chase's resourcefulness would come in very handy. When Chaplain Miller asked about Chase's faith, he found out he had a mixture of Catholic and Assembly of God, two very different ends of the spectrum. What Chaplain Miller got to experience was watching Chase's faith grow throughout this deployment.

The next two weeks were hectic, to say the least. A chaplain was selected to backfill for Chaplain Miller while he was gone, which meant he had to be shown the ropes in two weeks. So Chaplain Miller would spend his time between the unit in Conway and his office at Camp Robinson. There was no such thing as eight-hour days, and for these two weeks, he found himself, virtually every day, coming in late in the evening, getting as much time with the family as he could, getting some sleep, and doing it all over again the next day.

At the end of those two weeks, Chaplain Miller was granted 10 days of leave and just spent it with family. All of the staff and soldiers on this mission took the same. He rented a condo in Branson, Missouri, for the entire family plus his parents. He knew things were going to be hectic, and he wanted some quality time with those closest to him. They didn't plan a lot to do; the entire goal was just to spend time together and be together.

They got to the condo and when his parents arrived, his mother broke down. Nothing big, but she was a mom who knew her son was headed to a combat zone and acted like moms act, even though he was months from leaving. His dad was a lot quieter, but when he and Jeremy were alone, he asked some questions to make sure his son was ready to go. Jeremy remembers one conversation in particular.

It was late at night and the two were sitting on the balcony. Jeremy said, "I know you are worried. It isn't going to do you any good to be worried about me. I've got a job to do. I'll communicate as much as I can and I'm going call. I'm going to talk to you. But this is my job. This is what I'm supposed to do." Jeremy's dad asked, "Are you ready for this?" Jeremy answered, "Let's relate this to the carnival business that you've been in all of your life. On Saturday night in Jefferson City, Missouri, Cole County Fair, where are you? No, wait, I'll tell you where you are. You are sitting between the beer garden and the grandstand watching the crowd from the office. Why is that?" His dad said, "Because it's my job." To which Jeremy said, "And that's exactly why I'm going, because it is my job. My job is going to take me to watch troops, to minister to them, to cry and laugh with them, to hurt with them and to love on them. That's my job and this time that job just so happens to be in Afghanistan."

This is what servicemen across all of the branches of the military do. They go where they are sent. They work and they train to be able to execute the mission when it is given to them and execute it at the highest level possible, wherever that location may be. It does not exclude chaplains in any way. They are required to do the same thing each and every serviceman and woman is required to do.

The conversation continued. Jeremy's dad said, "But you don't have to go. You are taking care of the troops right here. You are choosing to go. You could back out of this." "No, no I can't. That's not what we do. There was never a question not to go. This..is..what..we..do! When it is time to serve, when it is time to put our money where our mouths are, we go. And I'm not going to ask somebody else to go in my place, even if I could. It is not what we do. It is why we serve." Dad asked, "Well, what kind of guns are you going to have?" Jeremy's answer didn't reassure him. "I won't have a weapon. My weapon is my Bible." "Then who's going to take care of you?" "Chase is going to take care of me. That's his job." "Is he a big guy?" "No, Dad, he's not, but he doesn't have to be. Dad, it's going to be all right. God has this under control."

During this ten-day period Jeremy was reading his Bible every morning, and one particular Scripture spoke to him, Proverbs 1:2 "***To know wisdom and instruction; to understand words of insight; to receive instruction in wise dealing, and righteousness, Justice and equity, to give prudence to the simple, and knowledge and discretion to the youth."*** (NIV) Those words struck home with him. To be wise and to understand that which is taught to you. To be able to receive and learn those things you need and then to have the ability to pass this same wisdom and knowledge on to those who need it, particularly those young troops being deployed. This stayed close to Chaplain Miller's heart.

During this time, it dawned on Chaplain Miller that he had grown up in the position he was now holding. He was only twenty-five years old when put into this position of leadership and was basically a kid. But through all of his experiences, he had grown into a mature adult capable of handling the hard and difficult situations. He knew, because of where he was going, he may be scared half of the time, but he also knew he was now equipped to handle what would be thrown at him. He was now prepared to dispense the wisdom, insight, and understanding that was needed now and that would be needed in the future.

The time in Branson came to an end. Jeremy's mom and dad headed back to the carnival and the Millers headed home.

Chapter 25

Off We Go

When they got home, Jeremy packed up his gear and reported for duty at Ft. Chaffee in Ft. Smith, Arkansas, for Pre-Mobilization Training (PMT) the next day. Using the experiences from his time at NTC, Chaplain Miller set up the chaplain office in the same building as the Tactical Operations Center (TOC) of the battalion. This put him closer to those he needed to be close to. There was a chapel but he knew he would be better utilized closer to the troops. The battalion began to do all of the training necessary to deploy, including extensive physical training. The goal of this training was to reduce the amount of time spent at Camp Shelby, Mississippi, when the battalion would come together and train the final time for real combat activities. PMT is designed to hone the skills that would become necessary. All of the firing ranges were put to use and, as mentioned, PT became extensive. When they eventually left Chaffee for Shelby, these troops would be ready for what came next.

There were a little less than a hundred troops at this stage of the deployment, and Chaplain Miller was able to get close to all of them. He showed up at the training, at the ranges, and especially at the PT. He began at this point to walk among them. Where they were, he was. He went to the chow hall and ate breakfast with them. They let him play with every toy they had. He fired the M-9 grenade launcher and the .50 caliber machine gun. He drove everything they would let him and he would do everything they did. He got to know his troops and they got to know him. He called this "a chaplain's dream." He began establishing great relationships with the senior NCOs because these guys were the true backbone of the unit and they were the ones who actually made things work. They were also the go-between for the troops to the leadership. Each night, he would do a Bible study and began working the troops through the book of Acts. By the end of the roughly two weeks at Chaffee, that Bible study was fairly well attended.

The unit came home for two days, giving them enough time to do their laundry and repack their bags and they were off again, this time to Camp Shelby in Hattiesburg, Mississippi, where they would pick up the rest of the battalion. Those days flew by, and two days later, they were preparing to get on the bus to actually leave. There wasn't a dry eye in the place, which is the way it is for every deployment. As a young cadet, Jeremy had loaded the bags and waved goodbye to his troops. Now as a captain, he was loaded on the bus with them. Someone else would lock up the armory. Someone else would make the notifications back home. His focus now shifted to the mission ahead of them. The buses pulled away, and there was dead silence for the next two hours until the buses pulled in to get lunch for the troops. About four hours later, they arrived at Camp Shelby, and then the fun began.

It is really hard to describe just how hot Camp Shelby is in the middle of the summer. All of the buildings are made of cinder blocks. Hurricane Isaac hit while they were there and after it moved through, the humidity moved up to levels that were unbelievable. In order to be at full strength, the battalion was supplemented with troops from Illinois and Rhode Island. Chaplain Miller had already immersed himself into the Arkansas troops, so when the other troops arrived, he began to do the same with them. This happened through time on the ranges, convoy operations, and other training exercises.

They kept a spreadsheet on the wall that detailed each activity, each day, and he was able to make sure he and Chase were where the troops were. Chaplain Miller made sure that each day the troops were invited to an end of the day PT session and he would lead those sessions. Between 20 and 50 people would show up for these sessions, and not always the same people. It was good for everybody because the chaplain was right there with them, leading them in physical workouts as well as providing emotional and spiritual support. These were so effective that they put a sign above Chaplain Miller's quarters that read, "Even Jesus loves burpees." For those who don't know what a burpee is, as Chaplain Miller has described it, "It is an exercise movement designed by Satan." You start in a standing position. You then throw yourself to the ground and perform a push-up allowing your chest and stomach to hit the ground. Then you push back up, landing in a squat position, then jump and clap your hands over your head. He believed it was called a burpee because that was the sound one would make after doing about a hundred of them in a row.

A typical day at Shelby for Chaplain Miller went something like this. Get up in the morning, shower, and prepare for the day. This usually involved a short Bible study and time spent in prayer. Chaplain Miller would then go to the dining facility and have breakfast with the troops. Once complete, he would head to the morning staff briefing and figure out who was going where. When the vehicles headed out, Chaplain Miller would go in one and Chase would go in another and they would head out with the troops. They would spend the day with the troops, eat lunch with them in the field, and return with them in the evening to the PT pit. Once that was completed, they would eat supper and start preparations for the coming day. Days blended together, and it was very easy to settle into a semblance of a routine. That was a good thing and a bad thing. Good thing was there wasn't a lot of variation. Bad thing was there was not a lot of variation and troops got comfortable. That might be OK at home, but definitely was not OK in a combat situation.

Chase proved to be invaluable during this time. When Chaplain Miller would come to the office after eating breakfast, he would find Chase studying the spreadsheet to see where the troops were going to be. He would make sure Chaplain Miller was going to where he could do the most good. He would treat Chaplain Miller like a weapon system and deploy it where it would be most effective. He was always able to procure transportation, whether that meant a ride with someone else or a vehicle from the motor pool. He was a good NCO taking care of his officer. When they were apart, he stayed in constant communication with Chaplain Miller about where they were and what was coming next. If a soldier needed something or was having an issue, he made sure Chaplain Miller was briefed before seeing that soldier. He kept a great pulse of the troops and was able to relay that continuously to Chaplain Miller.

Chaplain Miller did everything he could to blend and mesh with the soldiers. He'd had this opportunity on much smaller scales, but this was the first time he was able to do this with a unit of this size. Every single day, from the time he got up until the time he went to bed, he wanted to be a living example of the Gospel for the soldiers. This was his calling and he embraced it with everything he had. Working along those lines, he continued, through his Bible study sessions, teaching from the book of Acts.

A brief Bible lesson here about Acts. It is a book of beginnings, a history of the founding of churches, the initial steps in the formation of the Christian society in the different places visited by the apostles. It records a cycle of "representative events." All through the narrative, we see the ever-present, all-controlling power of the ever-living Savior. He works all and in all, spreading his truth among men by his Spirit and through his apostles. It gives an illustration of the power and working of the Gospel when preached among all nations, beginning at Jerusalem. It discusses the ministries of Peter and John and introduces the journeys and writings of Paul. It is the stepping stone to Paul's writings and teachings. It is a powerful book in the Bible and one in which Christians should be well versed.

Chaplain Miller involved the command staff for the battalion, and through this teaching, the staff became really close. His reasoning for doing this with the staff was because he had heard horror stories of staffs not working together and making deployments miserable, something he would later encounter on his first day in Afghanistan. He prayed this would not happen to his staff, and it was his goal to bring this group together as family. Teaching from the Book of Acts did exactly that.

The staff consisted of six officers, from lieutenants to a lieutenant colonel; S1, S3, S6; and the chaplain. A military staff is a group of officers and enlisted personnel who are responsible for the administrative, operational, and logistical needs of a unit. "S" designates the staff system of an organization, and the number following designates the area of expertise, with an officer normally filling the position. S1 designates Personnel/Manpower, S3 designates Operations/Training, and S6 designates Communications. Even though it is a member of the staff, the chaplain does not have an "S" designation. There were another nine enlisted individuals working with those officers. Those consisted of a command sergeant major, who worked directly with the commander and served as the liaison for the enlisted soldiers, down to an E-4. So all in all, counting Chaplain Miller and Chase, there were fifteen members of the staff element.

All of these soldiers would attend these Bible studies, all of varying religious backgrounds, including some who had never attended a Bible study. This was a phenomenal teaching opportunity for Chaplain Miller, and he used this time to bring the staff element together. He encouraged questions and would spend half of the time answering these questions. As the study continued, individuals from the other companies began hearing about it and began showing up for the study time.

At the end of August, just before the unit deployed to Afghanistan, they were given a pass to spend some of those last few days with their families. Beth, the kids, and an exchange student Beth had signed up to host while Jeremy was deployed, arrived in Mississippi. Nicoline Jensen literally became a part of the family while Jeremy was gone. She had just joined the family days before they headed to Mississippi, and this was Jeremy's first time to meet her and get to know her.

Nicoline came from a bad situation from her first host family. Upon arrival, her host family moved her into a trailer and she wasn't even allowed to live in the host family's residence. A friend of Beth's mother called her about this situation and inquired about moving Nicoline, and Beth took her in. This was God moving for the entire family. She turned out to be a great help with Beth as Beth's pregnancy advanced. She became so close to the family that to this day, they are still in contact and spend vacations together.

The three days went by quickly and it was time to say goodbye for the second time, which proved to be much harder. As Chaplain Miller described it, "When we left Arkansas to head to Mississippi, I knew I would be seeing them again soon. This time it was like ripping off a Band-Aid and making it bleed all over again. I knew when I took my last look at my family that it would be twelve months before I saw them again. It was a pain that is hard to describe. The first time, I was the one driving away. This time, I was standing in front of a tan cinder block building watching my family drive away with everything I held near and dear to my heart." As Beth tells it, she got out of Jeremy's sight and had to pull the car over because she couldn't see to drive through the tears, and there wasn't a dry eye in the car. It was a tough moment for everybody.

Not to sound hard here, but this is the life of soldiers, marines, sailors, and airmen in our armed forces. When they put their names on the line turning their lives over to the military, they are literally doing just that. At any moment, a GI can be tasked to leave anything and everything behind and move halfway around the world. It is hard on both the GI and their families. Troops have to function and do their mission while at the same time knowing loved ones left behind are taking care of things. Divorce is extremely high during these times because either one or both of the spouses just cannot function this way. But it is what they sign up to do and they do it willingly. It is an extremely serious decision to sign up to defend our country against all enemies and it may take the most extreme sacrifices to do so.

The next day, they departed Shelby and Chaplain Miller realized he had everything he was going to need to survive for a year in three bags; there was a duffel bag, a rolling bag, and a rucksack. He was apprehensive about getting there because all communication with the unit they were replacing did not sound good. They didn't communicate with each other, they were constantly fighting with each other, and their deployment had been one of misery. These communications moved Chaplain Miller to a point to where he said to himself, "I will not let this happen to us. We will not feel like this about one another." He steeled his heart and mind to this mission.

The buses pulled up and the unit began to load to head to Keesler AFB for departure. The staff was the last to load. As Chaplain Miller was closing up his office, he hung a chaplain's kit (consists of communion oils and other items he was not using) on the door with a note that read "I'm deploying. You need this more than I do." Just a little humor to lighten up what can be a really depressing time for everybody. This was the beginning of making sure to try to lighten the mood for everybody.

After putting his stuff on his bus, he had to finish the out-processing process and went back through the TOC. The out-processing is handled by individuals stationed at Camp Shelby. Going through each station, he came to one designated as "Chaplain." This was the first chaplain he had laid eyes on in nearly three months. Chaplain Miller asked who he was and the captain said he was one of the chaplains stationed on the post. Upon further discussion, Chaplain Miller was told they didn't even know he was there. He was also informed, had they known, he would have been assigned a bunch of additional duties that would have kept him from becoming so close with the troops in his unit. This was both a good thing and a bad thing. The good thing about them not knowing he was there was that it allowed Chaplain Miller to really get close to his folks. The bad thing is if he had really needed some training, he wouldn't have gotten it.

Upon arrival at Keesler AFB, the entire unit was placed in a large aircraft hangar for several hours. Another truism of the military: hurry up and wait. The group had to keep busy with something and so the bags were separated and placed in different areas to be loaded on pallets when the troops began to form up. The hangar had no chairs and a small bathroom. About the only thing to do was find a spot on the concrete and get a nap. At that time, the preferred method of listening to music was an iPod. An iPod could store a decent amount of music either as albums or as singles. As they began getting into formation, Chaplain Miller's iPod had a malfunction, leaving him with two albums. One was Credence Clearwater Revival's Greatest Hits and the other was a worship album from David Crowder. While these were two really good options, it severely limited his ability to zone out during the flight.

Once boarded, Chaplain Miller and Chase found themselves at the back of the plane. Chaplain Miller settled in and had just closed his eyes to get some sleep when he felt someone poking him. He opened his eyes and CSM Stain was standing there poking him with his finger to get him awake. He advised Chaplain Miller that he was sitting in the wrong seat and wanted him to come with him. As you will remember, a CSM had taken care of Chaplain Miller once before and, as luck would have it, this was the same sergeant major and yes, and he was moving the chaplain once again to the front of the aircraft. CSM Stain quickly became one of Chaplain Miller's favorite people. He knew the work that Chase and Chaplain Miller had been doing this entire time and how it took pressure off of the command from having to deal with some of those issues. He also knew from multiple experiences that the chaplain was going to be busy. He needed to rest now. When Chaplain Miller asked if he could give his seat to someone else, the CSM said "No, I want them angry and ready to fight if needed."

Chapter 26

Romania

Eight hours later, the flight landed in Romania. Why Romania? Well, this is the Army way and at times things are beyond explanation. But needless to say, the unit was now in Romania, where they would stay less than twenty-four hours. Upon arrival at the airport, the unit was trucked to a small military installation just outside of town. Chaplain Miller described it this way:

"The city was just like you see in the movies, with huge Soviet era buildings and architecture. Upon arrival at the base, we got into formation in front of this massive building that was clearly of Soviet design. It had several rows of windows that gave the impression of several floors inside the building. The S1 (Personnel) gave everybody a piece of paper and we were instructed to write our names on it.

"When we walked into the building, we realized it was all a façade. The entire building was one giant room. There were four stories of windows. A table was sitting in the middle of the room and it was manned by two people, one male and one female. They looked like they had walked in from a movie set with all of the stereotypes one sees of Soviet military administrative personnel. The man was a little bit overweight, starting to bald, and had a grizzled expression on his face. The woman was a little overweight and had way too much blue eye makeup on. Both were in Romanian army uniforms, which strongly resembled the Soviet uniform of the era.

"Each of them have a stamp and a stack of paper in front of them. The man yelled something in Romanian and an individual at the door told us to get into two lines, one in front of each person at the desk, and when finished, we were to leave through the back door. I happened to get in the lady's line. She had a large, bleach blonde, bouffant hairdo to go along with her massive blue eyeshadow and very tight uniform. She said to me, 'Military ID, please,' in a strong, Eastern European accent. I showed her my ID and she hit my piece of paper with a stamp as hard as she could. She hands it back to me and said, 'Welcome to Romania,' and I headed out the back door.

"Three hundred of us were standing around, still kind of speechless by what had just occurred. As I was standing there, I said, 'I'm waiting for the guy with the Adidas track suit with a muscle shirt and gold chains to come driving up'. I couldn't help but get the image of the guy in the movie Behind Enemy Lines out of my head. Again, I was trying to keep everybody light.

"We marched out to a hay field where our Quonset huts were, literally with chickens running around in the field. I kept saying 'There is going be a guy in a track suit show up sooner or later with rap music playing as loud as it can' and the troops thought I was a prophet because of what happened next.

"This truck pulled up with our bags and all we could hear was Beastie Boys playing. It was an old Beastie Boys tune and it was as loud as it could be. The guy stepped out of the truck and he had on a gold chain with a white tank top with his hairy chest showing and was wearing, you guessed it, a black Adidas track suit with Reebok shoes. He jumped out of the truck and yelled in a thick Eastern European accent, 'Welcome to Romania, my friends,' and said, 'Unload truck,' which as we found out was the only English he knew. He sat by the truck and picked his teeth with a piece of straw until the truck was unloaded, and then drove off. He lived up to every stereotype Americans have of Eastern European men. The troops were amazed and kept asking me, 'How did you know?' and I kept asking them, 'Don't you watch the movies?'"

Chapter 27

Afghanistan

The next morning, the unit ate and gathered back in the field. They were again picked up by trucks and transported back to the airfield for the flight to Afghanistan. They boarded C-17s, with the commander, CSM Stain, and Chaplain Miller being the last troops to board. The planes were arranged with regular hard-back seats on pallets and jump seats against the bulkhead. Several hours later, the aircraft began a corkscrew approach to Bagram airfield in Afghanistan. What a corkscrew approach entails is the flight is level, the landing gear drops, and the flaps are set. Then the aircraft rolls abruptly and begins spiraling downward like a slowly turning bit in a power drill. The only appropriate term for this maneuver is to call it a controlled crash. By the time one figures out what is happening, the wheels touch ground and the aircraft rolls to a safe stop on the runway. This is done in order to prevent the enemy from shooting the aircraft out of the sky. Those troops who had been on previous deployments knew what was coming, but those who had not had a rude awakening.

As the aircraft taxied along the runway, the ramp began to lower and the heat slammed into the interior of the aircraft. Welcome to Afghanistan. Mixed with the landing, the smell of jet fuel, and the heat, some individuals became nauseated but were able to quickly get over it. When the aircraft came to a halt, the troops were instructed to grab three bags and get off of the aircraft. They were told not to worry about whose bag was whose but just get three and get off of the flight-line.

Eyes had to readjust because everything went from the gray of Romania to the brown of Afghanistan. Buildings, ground, and just about everything else was a mud brown. Bagram is in a mountainous region but everything is still mud brown. Chaplain Miller's first thought as he looked at the mountains surrounding Bagram was of King David when, in Psalms, he was describing Jerusalem down in the low part and his enemies were surrounding him in the mountains. Then the thought came to him, "There's bad guys in those mountains, and they are probably watching us right now." The realization of being in a war zone quickly struck home.

The advance party for the deployment had been in-country for a couple of weeks, and one of their jobs was to make the arrival go smoothly. High on that list was to procure transportation for the battalion. Everybody marched with the three bags to an in-processing area. After being in-processed, they were able to begin acquiring the proper bags and luggage and get ready for their transportation to arrive. Being in the Army, there were none surprised when they were told there was no transportation and everybody would have to ruck to their destination. Bagram is a huge military installation and there are no short walks to anywhere. They humped it a couple of miles with their three bags each to a tent city which turned out to be the staging area. The tents were large, circus-style tents filled with bunks and were where the battalion staged.

The battalion was tasked with manning ten different locations, including Bagram. For the next ten days, different elements of the battalion would transition to their assigned locations from this staging area. Since the command structure was at Bagram, command moved to their permanent quarters that day. Chaplain Miller thought he had moved into a post-apocalyptic world. They were literally living in an area completely comprised of CONEX boxes and plywood huts that looked like an organized shanty town. A CONEX is a trailer-looking box that is usually seen aboard large shipping vessels. They are stacked on top of each other and moved to a tractor rig to be transported to wherever the material it contains is needed.

The military learned how to take a number of CONEX boxes, stack them together, and make living and working facilities. These can be multiple stories high. Chaplain Miller's lodging would not be as strong as a CONEX. His team would be assigned to a simple plywood hut with a tin roof and a door with a key pad. This was to be home for the next twelve months for the approximately thirty soldiers assigned to Bagram.

About an hour after arrival at their quarters, Chaplain Miller got to meet the man he was replacing. We will call him Chaplain A. In the two-and-a-half months prior to getting there, the only communication between the two was Chaplain Miller being advised to bring shower shoes. That was it. Chaplain Miller had a rule and it went something like this: You learn from following either good chaplains or bad chaplains. Unfortunately, this man fit in the latter category. When meeting, Chaplain Miller went to shake his hand, and it was a rather limp handshake. Being a southern man, Chaplain Miller was taught to always give a firm handshake; you could tell a lot about a man by the way he shook your hand. This was strike one.

When Chaplain Miller asked about sleeping quarters (he wanted to get rid of those bags he had been toting around for hours), Chaplain A told him he would take him to the area where all of the chaplains slept. Chaplain Miller was kind of surprised because he wanted to be where the troops were, but this man had all of his chaplains and CAs under one roof. When Chaplain Miller asked him what would happen if a rocket hit that building, Chaplain A's response was, "Not my problem. I'm out of here in a couple of days." Strike two.

Chaplain Miller and Chase knew this lodging would be temporary so they threw their stuff into the place Chaplain A set up for them. Come to find out, it was a transitory barracks where different people who are there for only a short time stay, and Chaplain A, his CA, and one other ministry team were the only permanent residents there. Chase discovered this CA had requested a transfer. He simply could not work with Chaplain A. Fortunately, he was able to help Chase get information they would never have gotten otherwise. Another example of the NCO communication pipeline.

Chaplain A took Chaplain Miller to what can best be described as a red plywood chapel and one of the oldest chapels in Afghanistan. Chaplain A stated he had not been allowed to preach in the chapel. Chaplain Miller was astounded that the chaplain he was relieving had never preached while in-country. Even more astounding was the statement he had not been "allowed" to preach there.

The office setup in the chapel was outstanding. Chaplain Miller's office was private only to him and only he had the code to allow access to his office. There was another common area where troops could gather and unwind. It had a popcorn machine, TVs and VCRs for movies, and a reading area. They also had an area where troops could make DVD recordings for their families. When they were completed, the DVDs were left with the chaplain and they would use the USO to mail them directly to the soldier's family. Right next door to that was a thirty-five seat chapel with musical instruments and everything needed to conduct a worship service. Upon seeing this, Chaplain Miller could see the potential for this building had not been utilized.

They moved to the office in the chapel and sat down to chat. Chaplain Miller was anxious to get the lay of the land and where to begin. What he got was a gripe session. Chaplain A started, "Let me tell you a couple of things that I'm really mad about. I've been running all over this country trying to get a Combat Action Badge and I haven't gotten one yet." The Combat Action Badge (CAB) is awarded to any soldier serving in an area where hostile fire pay is authorized and who is personally present and is actively engaging or being engaged by the enemy and is performing satisfactorily in accordance with the rules of engagement. In other words, Chaplain A had not been under enemy fire and had been "running all over the country" trying to get under fire just to earn a CAB. He went on to say, "So I'm telling you right now, if you want one, you'll just have to get the S1 to fill it out for you. That's the only way you are going to get one. All of my other chaplains have been to the locations and they've had mortars dropped on them. I show up for a week or two and nothing happens. I haven't been shot at once." Chaplain Miller's response was "Praise God you haven't been shot at." Chaplain A responded "I can't get a CAB if I don't get shot at and I wanted one out of this deployment." He was visibly upset that he had not been able to earn a CAB. Here's where this gets bad. He didn't want the CAB to show he had been in combat or anything like that. He simply wanted the CAB because he felt it would "enhance his chances for promotion." So, as Chaplain Miller realized, Chaplain A would only travel to his outstations (FOBs, COPs) when they came under fire. If they did not come under fire, he didn't leave Bagram. Strike three. Chaplain Miller came to the conclusion that he would listen to everything this guy had to say and do the opposite.

Chaplain Miller then asked him what the second thing was that had him so angry. "They won't approve the Bronze Star medal that I wrote for myself." At this point, Chaplain Miller had to exert a lot of self-control not to laugh out loud. Proper protocol is that awards are written by someone else, generally someone higher in the chain of command. Chaplain Miller asked, "You wrote your own citation for a Bronze Star?" Chaplain A explained, "I'm a staff officer in-country in a combat area. I've done my job and I should get a Bronze Star." The requirements for a Bronze Star are heroic achievement, heroic service, meritorious achievement, or meritorious service in a combat zone. Chaplain Miller could not help himself. "Bronze Stars are earned, not deserved," he stated out loud before he could stop himself. That's way more than three strikes. You're out!

During the entire conversation, not once did Chaplain A talk about how he did ministry and how he worked with the troops. He never talked about the guys at the FOBs and COPs or how he had led anybody in worship. All he did during the entire time they were together was complain about what he had not been able to get or what he deserved to happen for him. All in all, there was not one single experience anyone would want to emulate. At the end of the conversation, he informed Chaplain Miller they would be traveling to Kandahar the next morning.

There was, however, a silver lining to that day. The office in the chapel had a phone with the capability of calling the states, and Chaplain Miller took full advantage of that to call and talk to his family. After what he had been through that day, he was able to go to bed that night with good thoughts that can only come from visiting with those closest to you.

The next morning, the outgoing CA showed Chase how to make travel arrangements for travel within country. Contrary to popular belief, you don't just run out and jump on an airplane or helicopter and they take you where you want to go. These things have to be scheduled.

After making arrangements, they flew to Kandahar in a C-130 and headed off to their first travel experience in-country. This flight was much different than the C-17. There were no real seats, just jump seats against the bulkhead. A jump seat is a seat made from webbing that you open up and kind of wrap around your body and it takes your weight and keeps you off of the floor. If an individual doesn't want to sit in those seats, you find an area, clip yourself in, and get comfortable on the floor of the aircraft, which is what Chaplain Miller did. He got situated and as the aircraft began to taxi, he put his earphones in and turned on his broken iPod. As the plane lifted off, the rear ramp came down to give him a view of the lands around Afghanistan. As luck would have it, the first song he heard while traveling in this combat zone was Fortunate Son by Credence Clearwater Revival. While nearly synonymous with the military, the song's origins are anything but. Definitely ironic, he thought to himself as he watched the landscape slip by.

The one good thing about the flight was that he was with some soldiers from his unit headed to Kandahar for their tour. He was able to visit with them during the trip. However, Chaplain A managed to further cement his place in Chaplain Miller's mind of being a bad chaplain. About fifteen minutes into the flight, Chaplain A told Chaplain Miller to come up to the front of the aircraft because he's been able to procure him a seat in the cockpit. He tells Chaplain Miller "Some first sergeant thought he was going to give some specialist a reward and get to ride up there but I kicked him out. You know, rank has its privileges. I do this all the time." Chaplain Miller was incredulous. He declined the offer and went back to his spot in the back of the plane. Passing that first sergeant on his way back to the seat, the two shared a nod and the NCO let out a few expletives regarding Chaplain A.

Upon arriving in Kandahar, it was readily apparent that it was distinctly different from Bagram. There were no mountains, it was all desert, and it was even hotter. They humped over to the compound where the element of the 39th would be stationed and got organized. It was late in the evening, and the quarters were semi-permanent tents. The inside walls had been sprayed with expanding foam to provide some insulation, and there were plywood partitions creating sixteen separate sleeping areas. The beds were literally planks raised about a foot off of the ground (this helped keep out the scorpions) to hold a sleeping bag. That was it. After eating, they retired for the evening. Just about the time his head hit his waded up t-shirt he planned to use as a pillow, Chaplain Miller heard incoming rocket fire impacting on the other side of the base. True to form, Chaplain A jumped up and said, "C'mon! They are shooting at us!" and started running for the door to go where the rounds were impacting. Chaplain Miller yelled, "I've got it on good authority that you don't run towards the rocket fire." Chaplain A's reply was, "If I can get close enough while they are coming in, I can get my CAB." It was clear to everyone present that the rockets were impacting a mile or more away. Chaplain Miller just shook his head, told him to have a good time, and lay back down. An NCO yelled from the back to shut up and close the door so the group could get some sleep.

Now, don't get the wrong idea. This was Chaplain Miller's first time hearing and being near hostile file. Yeah, the impact zones were about a mile away but still, it was incoming enemy fire with the intent of doing harm. His adrenaline was flowing and his heart was racing at the first thud. The shots were the real deal. But there was no way he was going to get out of the tent with spray foam (by the way, not very bulletproof) and go to where rounds were impacting. He also trusted the fact that his fellow NCOs were calm and so he followed their lead.

To help bring his heart rate down, Chaplain Miller put his headphones on and turned on David Crowder and listened to worship music. It did the trick. He prayed at that point, "God, I don't know if this is what the entire deployment is going to be like, but I don't want to be wasteful with the ministry you've given me like I'm seeing out of this other guy. I'm not praying that I'm fruitful and leading revival with thousands being saved. I just want to do the job that you have me here for and honor that purpose." He fell asleep soon after and slept through the night. At some point, the very sweaty, very upset, and very unsuccessful Chaplain A came back and lay on his bed, mad that he missed out on a CAB. The NCO from the back gave a less than gentle reminder to shut up and go to bed.

The next morning, they headed out to breakfast and to check in on the troops. They went with the 39th troops as they took charge of their areas of responsibility and relieved the other troops in place. They stayed that night, and the next morning they caught a flight back to Bagram with those troops who were headed home after being there a year.

Chapter 28

Settling In

Over next few days, as the 39th slowly took control, Chaplain Miller was able to observe the staff they were replacing. The biggest observation of that staff was that they did not like one another. They despised their commander, their chaplain, and each other. This was the very thing Chaplain Miller had been praying would not happen to his guys, and it steeled his resolve to make sure that did not happen. Within days, the 39th was totally in charge of their areas of responsibility. The Tactical Operations Center (TOC) was an old Russian bunker, about 20 feet by 30 feet. This would be their workspace for the next twelve months. Their first full day in charge, the commander gathered them all together and briefed on what they would be doing. Chaplain Miller prayed over the staff and the work began.

The 39th was now responsible, in some capacity, for a total of ten sites from the very northern part of the country around Mazer-E Sharif to in and around Kandahar as well as very western FOBs like Bastion. Some were on large bases and others were small dots on a map in the middle of mountain passes. To get to all of these sites, travel would be by aircraft of some type: Blackhawks, puddle jumpers flown by civilian contractors, and C-130s. Of course, because they were the largest, C-130s were the preferred mode of transportation.

The group commander used the first couple of weeks to allow his onsite commanders to evaluate what they were seeing and what they thought needed to be done. Because of the experiences encountered from the group they relieved, very little of what they were told was considered. This proved to be a wise move by the command structure.

This assessment period also applied to Chaplain Miller. Not really putting a lot of faith into what he had been told, he started from scratch. He and Chase began taking inventory of what they had and what they could use. This would lead to a tremendous find, but we will discuss that in a little bit. Early in this evaluation process, the senior chaplain for Afghanistan came to him and said, "I'm going to be leaving country in the near future and I would like to sit down and do some mentorship with you. I would also like to share with you how I think you could succeed in this mission. In addition, I would also like to ask you to start preaching every Sunday morning." (Remember, the chaplain who was relieved stated he had not preached a single day while in-country.) "All of this will be in preparation for you to take over when I actually do leave the country, but be ready, because it won't be long."

Needless to say, Chaplain Miller was excited by this opportunity. Part of being given this chance to preach came from Chase establishing a good relationship with the senior chaplain's CA.

It was getting near the end of fighting season and combat contact was increasing. There are fighting seasons in Afghanistan. Once the snows start falling, it becomes very hard for the Taliban to keep their resupply system in place. The snows close the passes, making it extremely difficult to get supplies through. It is also allows the Taliban to do two things. First, those fighters from other countries go back to their homes for a while. Secondly, it also allows the Taliban time to gather replacements for those fighters who have been killed. Because we own the air and we are a superior fighting force, those losses can be significant in any direct confrontation, and those replacements generally come from other countries. It is tough to do this when the passes get snowed over. Also, it is tough to stay warm in their traditional clothing, whereas we have superior cold weather gear and can fight much more effectively in those conditions. Keep in mind this isn't the case in all of Afghanistan. In the desert region it is fighting season year round. Those stationed at Kandahar will attest to that. But out of the desert, things slow down significantly when the snows start.

Chapter 29

Ramp Ceremony

Around the beginning of October, the task force that Chaplain Miller was supporting (fortunately, not any from the 39th) lost three soldiers, all from direct contact with the Taliban. Shortly after those first three losses, there was a fourth soldier killed in contact with the Taliban, leading to Chaplain Miller's first ramp ceremony. The soldier was identified as SGT McPherson. He was not part of Chaplain Miller's group, but the chaplain for that group had just finished three ramp ceremonies in a row and, frankly, was kind of emotionally wrung out. So, they asked Chaplain Miller if he would conduct this ceremony.

The ramp ceremony is a ceremony to load a fallen soldier's body onto the plane for its final flight home. The chaplain leads this ceremony.

The chaplain first talks and prays with the squad who will carry their comrade to their final flight.

Everybody who is available at the unit and also at the airport stands in formation on either side of the aircraft, forming a pathway to the ramp. The aircraft is parked with its back ramp down and is completely empty. The leadership is in formation facing the ramp of the plane. A vehicle carrying the soldier pulls up and the chaplain leads the detail that will carry the casket to the plane. They remove the soldier from the vehicle and carry the flag-draped coffin in a slow marching movement. The formations are called to attention and they render salute. The chaplain leads this detail into the aircraft, as well as all members of the command element. The chaplain marches as far to the front of the aircraft as he can and does an about-face. The detail marches to within two or three steps from him and set the casket in place. Once in place, the detail stands at attention on each side of the casket. The command element take their places and then everybody that can fit into the aircraft enters and takes their places. In the loudest possible voice, the chaplain delivers the prayer and eulogy.

The prayer for SGT McPherson was, "Lord, we mourn today for the loss of SGT McPherson, our brother and friend. We come to You asking for comfort to his family at home and his family here. Yet while in our grief, we are reminded we are not a people without hope. We are people that place our faith in a powerful God, Savior, and Redeemer. To those gathered, I pray that you put a resolve in our hearts to continue to fight. We ask for strength now from You, Our Lord and Savior." After the prayer, Chaplain Miller read out of Psalms and delivered a few simple sentences.

As soon as that was over with, the detail saluted, prayed over the casket and exited through the egress door on the aircraft. One member of the detail, the soldier escorting his "brother" home, moved to the left of the chaplain. The command element came through, they saluted and exited through the egress door. Every person in the aircraft repeated this maneuver. The chaplain remained in place until everybody had exited the aircraft. He then prayed with the escort soldier, gathered all of the coins left by various soldiers, put them in a box, and gave them to the crew chief of the aircraft. The chaplain exited through the egress door and then began to do his job by counseling with those still milling around.

Chaplain Miller and Chase headed back to their quarters to try to get some rest because these events always take place at night. This was Chaplain Miller's first one, and he was beginning to come down from the adrenaline and actually started shaking a little bit. But as is always the case in combat zones, he was not given much time to relax. About six hours later, he was notified another soldier had been killed and he had to do it all over again the next night. It was symptomatic of what happens during a deployment. Emotions and feelings go from one extreme to the other. A soldier goes from rest to rush to rest, sometimes with only hours between each peak and valley. Within twenty-four hours, Chaplain Miller had gone from never having done a ramp ceremony to doing two.

When he got to the second soldier's group, he found the six men who would escort him on the plane eating as much as they could as fast as they could. He asked if they needed anything or if he could pray with them and they told him, "No, Chaplain, thanks, we are good right now." These six guys had the wildest, wooliest beards he had seen in-country because they were Special Forces. They asked him to sit and eat with them, and they began to tell him everything about the soldier's life. What they told him caused Chaplain Miller to scratch out the prayer he had prepared and create a new prayer based upon these stories.

This ceremony was exactly like the other but was at the same time altogether different, simply because it was a different soldier. It may be some of the same people but it was a different ceremony. This time when he stepped through the egress door, Chase met him at the bottom of the stairs. Chase had taken the time to make the rounds among many of the NCOs to check on them. Chaplain Miller asked, "Chase, you OK?" Chase looked up at him and said, "Chaplain, I can't do another one of these this close together. It's too soon and I'm all over the place. I don't even know these guys and it hurts." Chaplain Miller asked him, "You know this is our job, right?" and Chase replied to the affirmative. But then Chase asked, "How do you turn it off and on?" Chaplain Miller replied, "I don't turn it on and off. Chase, I just don't show it." Chase said, "Then I need to teach you how to play poker because with that type of face, you could be pretty good." They laughed and Chaplain Miller said, "Chase, if this happens to a soldier, we are not going to drop the ball. It might not be done to somebody else's standard but it will be done to ours and it will be done well." After all of the ceremony prep, counseling, and the wrap up, Chase was spent emotionally, physically, and spiritually, and unknown to him, Chaplain Miller was as well.

Chapter 30

New Quarters and a Surprise

Chaplain Miller and Chase had begun to question if this was how it was going to be all of the time. There had been five ramp ceremonies in 15 days on top of extensive travel to each outstation, and that was all they could see in the future. This was going to suck and they knew it. This was another instance of having to "embrace the suck," and they were wondering just how much of this suck would they have to embrace. Civilians don't understand this terminology, but GIs do. It's tough to explain, but understood among those who have served.

They were at the chow hall and the guys eating with them understood the chaplain and his assistant were trying to get their footing. So one of the captains asked, "When is our next Bible study, and are we still studying the same thing?" This one question took Chaplain Miller out of the lull he was feeling and brought him back to what he was doing. Yes, he had to deal with the dead but he also had to minister to the living. Again, another lesson why God had brought him to Afghanistan.

The next day, Chaplain Miller asked the commander if it would be OK for him and Chase to move into their barracks. Remember, they were living by themselves close to the chapel. They wanted to be closer to their unit and the troops to whom they would be ministering. The commander gave the thumbs up to this and they spent the next day, which happened to be their laundry day, moving to new quarters with members of their command.

After they got moved in, Chase and Chaplain Miller compared notes on their new living quarters. Miller said he could tell whether it was day or night just by looking through the cracks in their two-inch thick plywood hut quarters. It goes without saying this would be an issue once winter set in. The hut was painted red and had a few sand bags around it. The red paint was mixed with turpentine to keep the bugs from eating the place to the ground. There was gravel all over the place with a crawl space beneath the hut between the gravel and the floor, which meant anything crawling underneath the floor could be easily heard. The roof was sheet metal. Now, don't picture the quality sheet metal you can get at Lowe's or The Home Depot. No, this was extremely thin sheet metal that had probably been there forty years. Imagine a low budget chicken coop with a single AC unit to heat and cool. There was no insulation. Plastic sheeting had been installed underneath the roof so that when the rain or snow came, it would drip through the roof onto the plastic and channel to the outside.

After moving into their "new" quarters, the next day, the previously mentioned great find occurred. Chase came to Chaplain Miller all excited. "You have to come with me right now!" he exclaimed. When asked why, Chase said, "Remember when I had found a board with all of these codes on them? I thought they were for locks or something and there were numbers I was assuming were for CONEXes. So I went around to all of the CONEXes and I figured out the board. But let's not talk here." He looked around like he was looking for someone listening in on this conversation. When they got outside, Chase asked, "Chaplain, how do we get things around here?" "Well, we beg, borrow, or scrounge for them." "So we need something to barter with, right?"

"Absolutely!" "Well, I've got us something to barter with. Follow me."

There was a little maze of CONEX boxes between the barracks and huts, and they walked around between them. It was dark, so Chase shined his flashlight on one and said, "That one. Make sure nobody sees me open this door." Chaplain Miller asked, "Are we about to find a body or something?" Chase said, "No, it's something better!" "Better than a body?" the chaplain thought.

Chase opened the lock and sure enough it was a CONEX for the chaplain, and nobody had shared this information. It was completely full of Keurig K-cups. A CONEX full of coffee was a GI's dream. "Did you go through this?" Chaplain Miller asked. "No, I saw what it was and shut the door before anybody saw what was in it. I also removed the codes from that board and put them in our safe so only we have access to it."

They began going through the CONEX and found every flavor of coffee imaginable in a K-cup. It was absolutely packed and stacked floor to ceiling, front to back. This was the mother lode! They estimated there were over a well over a thousand boxes of cups. Everybody had a Keurig, and this was the ultimate bargaining chip. Chaplain Miller said, "From this point on, whenever we leave to go to a site, in your rucksack, you will put two or three boxes of these cups in it to give to the guys at the outposts. This was meant for troops, so we will get it to them."

Needless to say, from that point on, everybody looked forward to the chaplain coming for a visit. Not only did they give out boxes on the first trip, they took orders and kept notes on what brands the troops at each post preferred and took those on return trips. They found a GI's lifeline and made sure the troops were taken care of. This helped the troops in towers during long shifts, the quick reaction forces upon return, or the lowest of Privates who spent his day burning diesel and human waste in pits. Chase also discovered a second CONEX that was filled with all types of things. Everything from recreational equipment to movies to popcorn. These had never been opened by the previous chaplain.

Chaplain Miller just happened to know where he could get popcorn and taffy machines (remember, his folks own a carnival) and had his folks send them one of each. Starting in Bagram and expanding to other sites used the CONEX box's content to give their troops a place to come and rest, get popcorn and taffy, watch a movie and have a cup of their favorite coffee and forget about things for a while. This turned out to be a tremendous ministerial tool.

Chapter 31

Routine

Things settled into a routine. Basically, everything for Chaplain Miller's team was conducted at night while the larger task force had shifts covering all 24 hours, which resulted in Sundays being very long. On Saturday night, Chaplain Miller would try to get a few hours of sleep because on Sunday morning, he conducted an 8am and a 10am worship service. He would be at the chapel by 7am. The inside of the chapel was painted all white and had thick Afghan rugs and folding chairs. Nothing elaborate. It had a flat screen TV and some musical instruments in the small storage area, but no musicians. There was a small pulpit that Chaplain Miller had moved to the side because he just used a small music stand to preach from. This allowed him to put more chairs in the area to accommodate more people. Chase would handle the electronics and would put music on the TV screen for that part of the worship service, using such things as YouTube or IWorship. The service would start at 8am on the dot.

This service was not conducted for just the chaplain's personnel. There were individuals from every branch of the service in various states of dress. Some were in civilian clothes, some in workout clothes, and some in their battle dress. Unfortunately, only about 45 people could fit comfortably, 50 to 55 if they were jammed in. The service was conducted like any type of worship service anywhere else in the world. In fact, Chaplain Miller stated if you closed your eyes, you could imagine yourself sitting in a Sunday service in the states. While there for the year, Chaplain Miller worked through the book of Acts initially, and then on to

Ephesians to finish up. Both books of the Bible provided plenty of preaching material and he was able to work through them thoroughly. Looking back, Chaplain Miller spent a total of 144 days of his nearly 300 in-country preaching or teaching the Bible in some way. On all of the other days he would have found himself counseling or making the rounds among troops.

As soon as the 8am service was completed, everybody would leave, the chapel would be straightened up and it would start over again at 10am. Same amount of people, same type of people, and everything was repeated. In between services, someone from the staff would usually run to the chow hall and get Chaplain Miller and Chase something to eat and put it in their office. This allowed them to eat immediately following the 10am service and wind down a little bit. Just like at home, lunch was a good thing right after church.

After the 10am service and after eating, Chaplain Miller would set up shop in the chapel annex, which was next door to the chapel. For four hours, he would do one-on-one counseling and discipleship. Some of the counseling would be scheduled, and some would be walk-in traffic. Again, this counseling service didn't only apply to those assigned to the 39th. These counseling sessions involved, like the worship services, individuals from all branches, as well as civilian contractors.

The only break during this period of time was the weekly staff meeting, which was held around 4pm. Chaplain Miller would take the slides he had prepared earlier and go to the conference room of the TOC for the 39th. All of the remote locations would call in, which enabled everybody to be involved in this meeting.

Chaplain Miller would begin the staff meeting each week. "I was asked each week to begin the staff meeting. The commander or the CSM would say, 'Chaplain, lead us off. Give us the word of the day, pray for us, and tell us what you are doing.'" So, he led off each week in the same manner. After his briefing, he would listen to the rest of the staff, note changes where he needed to focus, and listen to the ten sites. He and Chase took notes and compared notes at the end. They used these notes to prioritize their travel schedule for the week once the counseling sessions were complete at 6pm. If there weren't any counseling sessions scheduled, he and Chase would try to lie down and get some rest, because the day wasn't even close to ending. They would get a bite to eat with some of the staff. Once chow was finished, they would head back to the chapel for a 7pm worship service. This was the earliest service for the guys who were hitting the night shift. They would perform the same services as the morning, two services back to back, ending somewhere around 9pm.

Then came the thing in which Chaplain Miller took the most pride. Immediately after the evening services, he moved to another location. During one of the staff meetings where the remote locations had called in, he came up with the idea of having all of the remote locations connected for a multi-site worship service. Many of the modern churches today are going with the idea of being one church in many locations. They have one location that streams the main pastor's sermon to each of the locations associated with their church. This allows all of the churches in several locations to receive the same message from the same pastor. Chaplain Miller thought to himself, "If they can stream in for the staff meeting, they can stream in for a sermon." He spoke to the individual who managed all of the IT stuff and lo, and behold, they were able to get each of the locations to stream in for his sermon on Sunday night. The technician who set all of this up for him thought this was the coolest thing ever. He checked around and discovered none of the previous chaplains had ever done this. Chaplain Miller became the first chaplain, as far as he knew, who streamed his sermon to each of the remote locations for which he was responsible. He took the technology of today and made it work for him. Not only was he able to speak to all of the remote locations, but anybody who had not made any of the previous four sermons during the day could also come to this one. Keep in mind Chaplain Miller did not preach five different sermons on Sunday. He had one sermon and he preached it five times. By the evening, he had it down pat and he was almost able to deliver it with no notes. The only thing the guys at the remote locations did not get was the song service.

This was Chaplain Miller's and Chase's Sunday for the entirety of the deployment, resulting in a long day. At the end of the multi-site sermon, they would call it a day and turn in to get some much needed rest. Most of the time this was good, but as everybody who's been in a combat zone knows, sometimes the bad guys like to shoot at night. Fortunately, those times were scattered and far between.

Monday morning, bright and early, Chase would get both of them up and lay out the travel itinerary for the week. Chaplain Miller called Chase his little Travelocity Travel Gnome. He had a gift for making sure they had flights to get into the locations they needed to get to. Of course, he used the K-cups to full advantage for this assignment. They then would discuss the travel schedule over breakfast and head out. The goal was to visit as many locations as they could on Monday and Tuesday and come back sometime in the afternoon on Wednesday.

During the deployment, it was estimated he and Chase flew over 100 sorties to the various units, resulting in their having traveled almost a third of their time while in-country.

When they would land at each location, they would talk to the staff and find out exactly what they needed to do while there. As long as they were at a location, they would make an effort to see everybody on each shift. This might require that they grab their rest when they could, but they made the effort to see as many troops as possible at each location on each trip. For example, upon arrival at a location, they would arrange to meet with the staff and find out any issues that may be prevalent. They would talk to the first sergeant to determine which enlisted soldiers needed to be seen and/or talked to. Then they would go around to those soldiers and walk among them.

Their primary focus was to be real, transparent, and ready to listen. Sometimes the soldiers needed to talk and work through some issues with someone other than their leadership structure. Sometimes the issues were significant and needed a chaplain's touch. Being among them might mean sitting in a gun tower while they were on duty. It might mean getting into a Humvee and going with them on convoy duty. It might mean sitting down in the chow hall and sharing a meal with them. But a chaplain has to be there in order to do his job, and Chaplain Miller made sure he was there with his troops. Sometimes, if the time permitted, he would hold a Bible study and made extra effort when that was requested.

Virtually every location made a worship location for them, whether it was a tent, a building, or even an office. When he was on location, it was emptied out and became their worship space. When the space wasn't available, they would find a space and make it a worship location. They used dining facilities, weight rooms, living quarters, parking lots, laundry facilities, you name it, and they made it work for worship.

By Wednesday, the goal was to be back to Bagram in time for the midweek staff meeting at the start of the evening shift. Chaplain Miller developed an amazing ability of being able to sleep while flying. Beth had given him a camo travel pillow, and he took this with him whenever he traveled. As soon as he got settled on whatever aircraft he was flying, he would lean back and go to sleep. They tried to keep their sleep patterns as constant as they could while traveling so when they arrived at Bagram, everything stayed the same. The neatest thing was when they came across C-130 aircrews that were stationed at Little Rock AFB, either active duty or the Air National Guard crews that called Little Rock home. They were able to talk about home with guys who actually knew what they were talking about.

One thing that always amazed Chaplain Miller, especially when flying in helicopters, was the countryside. The diversity was astounding. They would fly over the Hindu Kush mountains, down into river valleys, travel over desert terrain, over some of the bluest lakes one had ever seen. But always in the back of his mind he knew there were individuals in that countryside whose main mission in life was to kill him.

Upon arrival in Bagram on Wednesday, Chaplain Miller and Chase would go to the short staff meeting and then to supper, visiting with the staff as they ate. Upon completion of the meal, everyone would adjourn to the chapel for Bible study. This became one of the most fun nights each week. This was a time for them all (39th staff) to get together, study the Word of God, and kind of let their hair down. They would laugh, joke, and talk about what was going on in their lives and their families' lives. This was the driving force in this group becoming like family. Instead of hating each other like the group they replaced, these guys genuinely cared for each other, and the Wednesday night Bible study was a major part of that transition.

One of the big things they did was Trivia Night right after Bible study. They would do Stupid Trivia, which involved Chaplain Miller looking up the most inane stuff he could and they would have to guess what it was. They would do Music Trivia where he would find a song and start playing it until someone dinged in. Celebrity Birthday Trivia was a favorite. They kept a running score and it was ultra-competitive, in a really good way. It's kind of hard to argue when you are laughing with each other. On occasion, they would set up a screen and have movie night. At the end of these evenings, Chaplain Miller would put on his workout gear and go for a workout. This did two things. First, it helped keep him in shape and at a level of fitness to do the mission God had sent him on. Secondly, it allowed him to rest his mind. While working out, everything was forgotten and he was focused on the workout at hand.

Thursday morning, the travel would start again. Chaplain Miller and Chase would come back late Friday or Saturday, depending on how far they had decided to go to see their troops. Saturday was typically a rest and laundry day. Chaplain Miller caught up on his writings and put the finishing touches on his sermon for Sunday (if needed; most of the time it was done by Friday). Until the winter came, Saturday was shorts and t-shirt day, unless he was going to the local bazaar or to the Morale Welfare and Recreation (MWR) area. It was his day, and he used it to rest and recharge because he knew Sunday was coming.

Chapter 32
October

By October, everybody had settled into what is called a "battle rhythm". The dictionary defines this as, "A deliberate daily cycle of command, staff, and unit activities intended to synchronize current and future operations." What this means is everybody had settled in to their mission routine. Communication protocols had been set and were being readily used. Command elements were functioning smoothly, and information was being gathered and processed, and command decisions were being made from this information. Soldiers were doing their jobs every day and doing them well.

The same thing applied to Chaplain Miller. He used the thing that God said to Adam about keeping and working the garden (Genesis 2:15). He looked for joy in the little things. He remembered a flight to Kabul as a perfect example of finding joy in the little things. Chaplain Miller and Chase caught a Blackhawk helicopter to visit the base. Their helicopter was part of a flight of four. As they traveled from Bagram, they initially were flying "nap of the earth." For those not familiar with flying jargon, this means they were flying very close to the ground. It is done to avoid detection and being attacked in a high threat environment. The first part of the journey took them over desert terrain. As people who have flown will tell you, from the air everything looks peaceful. There is none of the street noise or activity one experiences when on the ground. Also, flying the way they were, an individual also gets the illusion they could reach out and touch things as they travel by them.

As they traveled that day, Chaplain Miller experienced a moment of peace by observing the landscape. The desert can actually be a very beautiful place and isn't always like the perception we have. It is not a flat featureless terrain. There are hills and valleys, different rock formations, and different colorations of the ground and hills. Chaplain Miller marveled at how beautiful things were. As they approached the mountain range, his marvel grew. Lush landscape began to appear alongside stream and river beds. As they continued to gain altitude to go over the mountains, there was a gorgeous waterfall and the beauty was amazing. As they got over the mountain, they passed over this amazing turquoise blue lake that sparkled in the sun, and there was a person sitting in an inner tube, fishing. Miller was a little taken aback by the contrast from what he was seeing from the air and what was happening on the ground. From the air it was a beautiful country and from the ground it was a war zone.

Things were always brought back to reality when they landed because it was time to go back to work. This brought about its own excitement. The soldiers at all of the outlying bases didn't function with a chaplain readily available. So when they landed and the troops on the ground were made aware there was a chaplain arriving, things got a little extra busy. The first sergeant usually met the aircraft and would inform both the chaplain and Chase of what soldiers they needed to talk to. The chaplain arriving was never a bad thing. First, as discussed earlier, it meant coffee, which is like life to a GI. It also meant mail and any packages from home. The chaplain arriving was a little like Christmas every time he showed up. The two weren't specifically mail carriers, but since Bagram was the main hub, a stop by the postal room could get mail to an outstation fast via the chaplain.

Chaplain Miller and Chase, unless they were talking privately to a soldier, never stayed in shelters. They would go and sit with the troops in the gun towers, manning the firing positions, or go and find a game of cards to sit and visit. This was how Chaplain Miller took on the work of ministry. It wasn't by sitting around in an office waiting for someone to find the time to come and talk with him. By walking among the troops, spending time with them where they were, he was able to truly minister to his troops. It was like Jesus. He went to where the people were and did not wait for them to come to him. Even though he was nowhere in their chain of command, they were his troops. He felt that God had given him these guys to minister to and love. This was his church.

In most instances, unless the air travel dictated otherwise, they would spend the night with the troops and leave the next morning. It was this type of commitment that earned Chaplain Miller his Bronze Star. He did his job, but he did his job in a way that was so much different than some of the chaplains who had come before him. He didn't do anything particularly heroic or daring but on the other side of the coin, he didn't sit around waiting for things to happen. He made sure the troops knew who he was, not because someone had told them, but because they saw it first hand at least every other week.

This was not to, in any way, diminish the role Chase shared with Chaplain Miller. As Chaplain Miller said numerous times, "A chaplain is a weapons system. A good chaplain's assistant knows how to use that weapons system." Chase was the liaison between the troops and Chaplain Miller. Chase would coordinate all of the trips. He would deal with all of the first sergeants. He would organize the visits and make sure Chaplain Miller was where he was supposed to be when he was supposed to be there.

Chase developed a close friendship with a first sergeant from Rhode Island named Jay Caetano. Jay was a huge New

England Patriots fan, as was Chase. This was tough on Chase because one cannot find Patriots gear in Arkansas. However, that was definitely not the case with Jay, and he showered Chase with Patriots gear. Chaplain Miller was pretty sure Jay got a little more than his fair share of the coffee disbursement as well.

Jay was stationed in Kandahar. When Chaplain Miller and Chase would arrive, Chaplain Miller would meet with the company commander, and usually they would go for a workout. It was the place where he could get the commander to really talk more easily. Chase and Jay would go eat and share what they needed to share with each other. When the commander and Chaplain Miller would return, Chase would already have the lay of the land and who Chaplain Miller needed to go see. There is a military truism that goes something like this: A unit is only as good as it'\s NCOs, and if it has great NCOs, it will be a great unit. The best officers surround themselves with the best NCOs. Jay and Chase were good NCOs and because of that, it freed up their officers to take care of the matters to which they needed to tend. Chaplain Miller depended upon and relied upon Chase, and without him, Chaplain Miller's job would have been much, much harder.

One particular trip brought home the fact they were in a combat zone. On the return flight from the previously discussed trip to Kabul, there was an incident. They left Kabul around 2am. It was October, and the nights were definitely beginning to cool. Chaplain Miller and Chase were on the flight line as the Blackhawks arrived and quickly boarded to head back to Bagram. The engines never slowed down, and from the time the copters arrived to when they took off again was a matter of seconds.

Everybody put on their NVGs (night vision goggles) and were just settling in when the mission changed. The helicopters began a circling pattern and stayed that way for about thirty minutes. When Chaplain Miller looked out, he could see a convoy on the road spread out in a defensive formation. They had taken fire and were trying to locate the position of firing from the Taliban. The Blackhawks were providing air support and cover for this convoy. They circled and searched for the enemy, and one actually received small arms fire as it approached a mountain. The door gunner/crew chief notified everybody of what was happening. The cool thing about NVG was that Chaplain Miller and Chase could see everything that was happening on the ground from the helicopter. After about thirty minutes, the convoy gave the all clear and everybody went back to their original mission. It was a reminder that combat action could break out at any time and any place.

Chapter 33

November and a New Arrival

The month of November was a banner month for the wives to have babies. You could almost pinpoint where this all started. There was a break before the troops headed to Mississippi for their final training, and nine months later, babies began to arrive. Chaplain Miller was not exempt from this phenomenon as Beth was due to deliver their youngest daughter sometime in November. Unfortunately for the troops, the policy is that they cannot go home due to the birth of a child.

To some, that may sound cruel, but if every soldier was allowed to go home from a combat zone because his wife was having a baby, it would be a madhouse trying to keep stations manned appropriately. The logistics would be a nightmare. This doesn't mean the military is heartless. There are instances where a military member is allowed to leave. The death of an immediate family member or a serious accident or critical illness of an immediate family member would be instances when a military member would be allowed to travel home. Those were always tough notifications for Chaplain Miller to make, and he tried to make sure he could make as many of those as possible for his troops. But the birth of a baby was usually done through the miracle that is the Internet and wireless communication.

Chaplain Miller made all sorts of plans to be ready for the birth of his new daughter. He did not travel the week Beth was due. He gave her numbers to call if she could not reach him directly. There was a designated computer where Chaplain Miller would have priority once he was notified she had gone into labor. The week before the baby was born, he was a mess. He called or Internet chatted with his wife numerous times during the day, wanting to make sure he was up to speed on everything. The only thing was, not a single plan he put into place for this birth was put into action. He lived a military truism: All plans are great until the first round is fired.

On the evening of Nov 13, Chaplain Miller checked his email and texts and there was nothing for him on either one. He had some time to kill so he decided to go for a workout. He had an extra-long workout and did something that he will remember all of his days, but not for the reasons one might think. He had a dead lift of 525 lbs., which was his personal best. He got back from his workout and again checked his email, his texts, and with the guys in the TOC, and there was nothing from home. So he went to eat. He returned to his quarters, did his Bible study and readings, and went to sleep.

What he was not aware of at the time was Beth had gone into labor. She couldn't reach him so she tried to call all of the numbers he gave her to call. Unfortunately, the phone system at Bagram was down at that time and she was not able to get through. She sent emails to him but he didn't see them because he was asleep. Afghanistan is 9.5 hours ahead of Greenbrier, Arkansas. This is why Chaplain Miller was asleep while his wife was giving birth to their daughter. When he awoke and checked his email, his inbox was flooded. However, it was the last email that stood out. It was from Beth and read, "Call me and you can meet your child!" Chaplain Miller ran to the phone and called his wife. When she answered, he remembers her saying "I've had the baby. Where have you been?" He answered he was asleep and she asked, "Where were you before that?" and thus brought about an answer that will live in infamy. When asked, Chaplain Miller replied, "I was in the gym and I got a 525-

pound dead lift." To say this was a really dumb thing to say would be an understatement. When the silence broke, the reply hit him square in the face. "I don't care. I just had a baby." Just goes to prove that even the best of men can be really moronic at times.

Chaplain Miller ended that conversation and ran to the computer to dial in to Beth. Fortunately, they had the clearest reception of the entire deployment right at that point. Beth still was not a happy camper because she simply wasn't able to reach him, but Chaplain Miller was able to soothe her feelings a little to get the entire story.

When Beth went into labor, things began to happen. First, of course, she could not get in touch with her husband. Their exchange student went to the hospital with Beth and was in the hospital room with her. When things started happening, the exchange student fainted and hit her chin on the floor, causing a gash that had blood gushing everywhere. She still has a scar on her chin to help her remember that day. It was chaotic. When he dialed in to see his wife and child, his whole family was there to remind him he missed it. For the next couple of weeks, he was the most apologetic husband and did everything in his power to try to make things right from half a world away.

However, there is a bright side to this adventure. When he is now sitting with his soldiers listening to them express their grief and anger about not being there with their wives, Chaplain Miller can truly commiserate with them and have a funny story of his own to tell of daughter's birth. This is another example of "embracing the suck." Things back home are going to be missed. There will be times when nothing goes right. It sucks, but it is part of the mission and the job.

So embrace it, know it is going to happen, and move on. Embrace the suck.

Thanksgiving was spent at Camp Scorpion near Kabul, and the troops feasted on roasted goat. Chaplain Miller observed how the shepherd slaughtered the goat and prepared it for cooking, and watched it being cooked over a period of two days. He and Chase decided to forego consuming any of the goat on Thanksgiving Day and they ate local bread and vegetables. This proved to be a wise decision because everybody who ate the goat got sick.

Chapter 34
Jalalabad Attack

On 30 November, they headed to Jalalabad. Part of the 39th BSTB unit makeup stationed at J'bad was a contingent of soldiers from the 169th Military Police Battalion of the Rhode Island National Guard providing base security. They would man checkpoints and gun towers surrounding the base. Occasionally, they would provide security for various convoys. They were a great group of soldiers to visit with, and Chaplain Miller loved to work out with these troops.

Upon arrival at Jalalabad, Chase and Chaplain Miller began to make their rounds and received the usual offers of playing video games, cards, or some other activity the troops had scheduled when their shifts ended. One of the guys he spent the evening hours with was Quinn Rogan. Quinn had received his Christmas presents from his family and really wanted to open them. He was like a little kid at Christmas. Chaplain Miller convinced him he should wait the twenty-five days until Christmas Day.

Chaplain Miller and Quinn had piled into a room with a couple of other soldiers and were playing a first person shooter game into the wee hours of the morning. During times like this there is no rank or position. Trash talking each other was not only expected but encouraged, and Chaplain Miller more than held his own in this regard. He was probably the oldest guy in the room by eight to ten years and was clearly the worst at video games. These moments weren't held because he loved video games. Instead, soldiers would typically follow him outside when he took a break and share with him issues they were dealing with back home. The video games were just an avenue to show his troops he was willing to connect with them.

Later that evening, he visited some troops in a gun tower. One soldier was always on the outside of the tower with at least two or three others on the inside. They would rotate standing watch on the outside of the tower. The walk to the towers took him over dusty rocks and powder. The dust was thick and made walking difficult in the area the troops referred to as no-man's land. It was a space between the base's furthest exterior walls and its second line of walls. This particular tower was on a major thoroughfare with five lanes of traffic. That entire side of the base at Jalalabad was paralleled by this stretch of highway. People were able to walk right next to the primary exterior wall of the base. He climbed the narrow staircase and greeted the troops.

CSM Stain was with the chaplain that night, and he gave the same talk to those troops in the tower that he always gave. "You have to wear your protective gear at all times. Your gloves, your glasses, your headgear, don't take them off because you think everything is going OK. Expect an attack at any minute." He proved prophetic with this talk.

The next morning, 1 December, they left J'bad and headed back to Bagram. Goodbyes were said and they headed off in a Blackhawk helicopter. Chaplain Miller remembered this being one of his best visits to date. Because they worked most of the night, Chaplain Miller and Chase headed to their hut and their beds. They slept through until the morning of 2 December. Chaplain Miller was barely awake when the commander began pounding on all of the doors yelling that everybody needed to get up because "they had to go." He jerked Chaplain Miller's door open and said, "We gotta go! They are hitting our guys in J'bad!"

This was the sequence of events. The soldiers spent the early morning doing their pre-checks after shift change. The soldier out front would be observing his sector while breaking up groups of people who would try and yell or talk up to them. A simple, "Imshkay" - Afghan for "keep moving" - was given, and the crowd moved away. Without warning, a car bomb exploded 100 feet or less from the tower. The target was the next tower over and the blast had just killed three Afghan soldiers. The blast shook the walls across the small base and soon everyone was under the realization that an attack had begun.

SGT Rogan had his team on the radio while he and another solider scanned the area for possible threats. The biggest threat was a car that had stopped in their area but was either not part of the attack or a failed car bomb. Soon, the pair ignored the car as enemy rounds began to impact around their tower fortification. The pair spotted the muzzle flashes coming from a construction yard across the street and immediately returned fire. Enemy targets began to pop up all over their sectors of fire before Rogan heard a man appear near the car yelling "Allahu Akbar." Before he could adjust fire on the man, a rocket-propelled grenade slammed into the tower. Then the car exploded, and the two soldiers bore the brunt of the blast. Miraculously, they were alive. The others in the tower were unconscious, but alive. With hearts pounding, they got back to their weapons and continued their defense as reinforcements began to fill the small tower. Among those was SGT Mike Church, SSG Steven Moretti, and SGT Carlos Zapata. All three were tough soldiers who had been sprinting to their people with additional ammo and weapons since the first blast.

As wounded personnel were evacuated, the team kept up the fight. Soon enemy mortar fire and additional waves of suicide bombers attempted to blow a hole in the wall or at least silence the machine guns on the tower. Hundreds of rounds impacting the tower paled in comparison to the rate of fire pouring over the construction yard and other areas across the sectors of fire. Without warning, a blast wave swept over the tower. Rogan recalled feeling the heat and seeing the orange cloud of fire sweep over him. When he woke up, he was at the casualty collection point and in pain. His hand and eye were severely damaged. Others on the team were mildly injured, but Rogan's was the worst. With his functional eye, he could see a hole in his hand and a blood stained uniform. He fought to stay alert and could still hear the fighting going on near his position. Waves of Taliban fighters and additional VBIEDs (vehicle-borne improvised explosive devices) hit other parts of the base as well.

Upon first contact, the base reported the attack, and Chaplain Miller along with the command team had begun to move. He grabbed his gear and headed out the door. Chaplain Miller, Chase, and parts of the command staff boarded a Blackhawk that was already spinning up and they headed to Jalalabad. The main thoughts running through Chaplain Miller's head were how, less than 36 hours ago, he was messing with these guys and was standing exactly where they were standing. Things got personal for him very quickly.

Upon arrival at Jalalabad, Chaplain Miller noticed the change in the posture of Chase, the commander, and the CSM. They had all been on the previous deployments to Iraq and they knew how to handle themselves in a combat situation. All three began moving with a purpose. Not having been in this situation before, Chaplain Miller just mimicked their behavior. He figured since they had been through this already, they knew exactly what they were doing and mimicking them would not be a bad idea. The primary difference was they were all armed and ready. Chaplain Miller did not carry weapons. But he knew he had soldiers who needed him as quickly as possible, and that became his primary mission.

By the time they landed, the attack was still underway but the enemy was losing steam. A few mortar rounds only alerted the quick reaction force (QRF) to the rooting out of enemy forces hiding on the base's edges. Chaplain Miller made his way to the casualty collection point to begin his work with the wounded. The first person Chaplain Miller ran into was SGT Church who had blood on his face and uniform, but he was still giving directions and orders, trying to keep things organized. Chaplain Miller noticed something out of place and called SGT Church over to him. When Church got there, Chaplain Miller saw it was a small metal piece near his cheek. He pointed it out to SGT Church and they got it out and wrapped it up in a piece of cloth. Chaplain Miller said, "You are going to want to keep this. Don't let the doctors take it away from you."

SGT Church asked for his troops to be prayed for on the spot, which Chaplain Miller was happy to do for the warrior. SGT Church stuck the metal shard in one of his pockets and walked toward the medics to get cleared.

Having finished with the wounded troops at the collection point after an hour, Chaplain Miller walked back to the tower where he had been 36 hours earlier. He saw the blast damage and the blood from the wounded soldiers. There was busted glass, concrete, and shell casings everywhere, indicating that even though wounded, his soldiers had done their job. The smell of the still smoldering car on the other side of the wall was nauseating - a mixture of burning oil, metal, and flesh. The bodies of the enemy fighters were scattered on the other side of the road and he imagined the smell indicated several more had been in the smoking metal piles that use to be a car at the tower's edge.

The next person Chaplain Miller saw was SPC Loren. The best way to describe her condition was rattled. She did not fully regain consciousness until after she had been evacuated from the tower, and when Chaplain Miller saw her, she was still not even close to recovered. She had no recollection after the blast. As he talked with her, somebody came to get Chaplain Miller and said he was needed at the hospital back in Bagram. Before he left, he was asked to speak with a nurse at the hospital. She asked if he was the chaplain and after affirming that, she said, "Quinn just woke up and he wants to know where you are."

SGT Rogan was in and out of consciousness during this time but he does remember two things. The first is, in his words, a really cute nurse treating him in Jalalabad and asking her for her phone number. (A GI. What can you say?) The second thing he remembered was waking up, finding out he was at Bagram, and remembering that Chaplain Miller lived there. He began badgering everybody to get the chaplain to his hospital room. He didn't want the hospital chaplain, he wanted his chaplain.

Chaplain Miller knew he had to leave but he needed to get around and see the other guys. One place he went was to Rogan's room and got as much of his stuff as he could carry in a pack, including all of the Christmas presents his family had sent. Jeremy said his goodbyes and jumped on a bird headed back to Bagram with wounded troops while the command team stayed to assess repairs. When he got to the hospital, he went directly to Rogan's room and sat there until his soldier regained consciousness after surgery. Rogan rolled over and looked at him with his good eye and asked, "How you doing?" They laughed about that.

In the end, Rogan lost one eye and most of the use of his left hand at that time. Chaplain Miller spent a couple of days in and out of the hospital making visits. On the second day, Rogan noticed that Chaplain Miller had brought his Christmas presents and asked again, "Can I open these?" This time, Chaplain Miller said yes and Rogan held up his hand and said, "Can you open them?" and they laughed again. The very first present they opened up was a video game. "Really???" Rogan said. "One eye and no hands and I get a video game." They both broke up.

The thought ran through Chaplain Miller's mind at this time, "This is what it means to be a chaplain. When your soldiers need you, they know you are going to show up." This emphasized to him why he was a chaplain and not a pastor. He knows pastors go through things but maybe not to the extent chaplains do with their troops. To this day, on 2 December, Chaplain Miller calls or messages Rogan and wishes him a "Happy to Be Alive" day. SPC Loren and SGT Rogan both eventually got out of the military, but SGT Church is still a member of the National Guard. Fortunately for this deployment, this was the closest they came to losing one of their soldiers, even though elements to which they were assigned lost soldiers of their own.

Chapter 35

December

The Taliban fighters consisted of many foreign fighters as well as Afghans who were fighting alongside them. These foreign fighters came from numerous countries around the Middle East: Pakistan, Iraq, Iran, and Syria, just to name a few. They had to be equipped, and in a lot of cases, came only with what they could carry. This meant they had to be resupplied on a regular basis and did not have the cold weather gear necessary to fight in the harsh winter conditions of Afghanistan.

When winter hit, the fighting season, especially in the mountainous areas, came to a halt. Again, this didn't mean all combat operations went away because the Taliban always did something to keep our troops on their toes, but primarily. Any large scale operations were put on hold. They used this period of time to resupply and reinforce their numbers because, in most cases, when they went directly against the American forces, they lost and, in some cases, lost big. They didn't have the steady resupply of troops, arms, and equipment the Americans were able to bring into country on a regular basis. So, this pause in the fighting for the winter months allowed them to bring in new fighters and equipment. They are very efficient in using social media to advertise and put out propaganda to resupply their numbers and financing. For example, they put out on Twitter (which our side keeps a close eye on) that they were going to rain fire down on the infidels, but the reality of it was they would fire one or two rockets that were ineffective. But by showing the rockets being fired, it showed those interested that they were doing something against the Americans.

The attack on Jalalabad was part of a coordinated last effort around the entire country and after this attack, things slowed down in a hurry. The Afghans took their last big shot and then retired for the winter. They were well aware of American holidays, which meant on those days, one could expect mortar or rocket attacks, which indeed happened on Christmas Eve and Christmas Day. Very small effort, but it did remind the Americans they were still around. The window of the pause of fighting season lasted from mid-December to around the end of February. This meant Chaplain Miller's mission changed.

The troops suddenly found themselves inactive, which might not be a good thing. They either were preparing for the fighting to begin or were fighting boredom. The biggest foe was fighting the weather, either to get to the troops or basically do anything. Weather resulted in travel being curtailed significantly and Chaplain Miller being stuck at Bagram for extended periods of time. He fought the same thing everybody else fought, boredom. By being stuck and not able to travel, there was only so much he could do. He could only work out so much or visit places around the base so often. This brought home to him exactly what the troops were experiencing and helped him deal with their problems more effectively.

One positive note from this period of time was that Chaplain Miller was able to spend increased time in ministry with the staff and those at Bagram. Before this time, they were only part of his rotation. However, now that he was stuck there, he began a serious study with those troops in the book of Ephesians. This more intense study with the troops led people to begin to question their own standing with Christ, and several individuals became born again Christians due to this study and the time Chaplain Miller spent with them.

Even though he was to provide spiritual support to the troops and staff, this was always the end goal, to have troops find Christ and expand their relationship with Him.

The first instance of this was a young female soldier who Chaplain Miller knew had been sitting at the back of the room during his Bible study time and didn't miss a day. Just before Christmas, she came to him and said she had been listening to what he was saying and preaching and she wanted to become a believer, a child of God. This wasn't due to any threat of losing her life or fearing she might die any day on a mission. She heard what was being preached and talked about and wanted to become part of it. She was his first baptism while in-country and this didn't happen easily.

The baptistery was at another chapel away from their task force. It was outside and frozen solid. At Chaplain Miller's request, the chaplain supporting that area broke up the ice, put in fresh water and baptized the soldier. People gathered around outside between services on a Sunday morning and watched her being baptized. They had to break up the ice that had formed on the surface just before she jumped in. Steam was rolling off of her head because of the cold, making her look like one of those football players sitting on the bench when it's really cold. After her baptism, she slipped into a pair of sandals and sprinted about fifty feet to the chapel to change and got into some warm, dry clothes. This turned out to be the only baptism Chaplain Miller was able to be a part of. A number of his troops who came to Christ during this time were baptized in another area of the base simply because Chaplain Miller wasn't around due to his travels. He would rather they followed in believer's baptism rather than wait for him to cycle back around. If they wanted it done, he found a chaplain in the area who could handle things as long as it wasn't one of his small outstations.

Chaplain Miller firmly believes that God is at work in Afghanistan, and there are Christians scattered throughout the country. He met his first evidence of that in December at the bazaar on the base. Some soldiers asked Chaplain Miller to go with them to the bazaar, which was a hodgepodge of shops selling everything from knives and swords to scarves and women's wear. It is very similar to huge flea markets encountered in the states, just with different types of items for sale. There was snow on the ground, so no one lingered outside for long. All of the guys told him he needed to go into the scarf shop because he would get a good deal and they would get a free scarf for bringing in a friend. Chase, of course, was with him because Chaplain Miller seldom got far from Chase's view.

The proprietor of the shop was dressed in the traditional garb of most Afghanis. He noticed Chaplain Miller and pointed at the cross on his uniform and asked if he was a chaplain. He asked Chaplain Miller if it was his first time in the shop and Chaplain Miller acknowledged it was. At this point, Chaplain Miller noticed that all of the other guys, Chase included, had stepped outside to look at the items hanging on the wall and he was all alone with this man. The man turned his back to Chaplain Miller and began to look around the room. The thought running through Chaplain Miller's mind was this guy knows I'm a chaplain and we are all alone in this room. Kind of made him a little nervous, to say the least. This went along with the assumption this guy was a Muslim, not a Christian. Luckily, this was a wrong assumption.

Keeping his back to Chaplain Miller, the man said "You know, God is a friend of mine." This was not the normal conversation of a man of the Muslim faith. Chaplain Miller said, "Excuse me, sir?" and the man responded, "God has been a good friend of mine for many years." Now Chaplain Miller understood why the man made sure nobody else was in the shop. He turned and grinned, pointing to the cross on Chaplain Miller's uniform and then at himself and then gave a thumbs up gesture. Chaplain Miller grinned as well and said, "I understand." He made the motion that he could not talk anymore, but he was telling Chaplain Miller in his own way that he was a believer. He knew he couldn't say any more and right at that moment, a number of people entered the shop and the conversation was finished. It gave Chaplain Miller the understanding that because people were dressed a certain way doesn't mean they are all the same. Men and women wore what men and women wore in Afghanistan, and unfortunately, we have become conditioned to think anybody dressed this way was in lockstep with the Taliban or Muslim extremists.

This is the major issue with making assumptions about people simply by how they look. This man was a Christian, but with certainty, when people looked at him, they saw a man of Muslim faith because this is how they have been conditioned to think. Unfortunately, this doesn't happen just in Afghanistan. This happens everywhere, including the United States, which is sad, but true.

Chaplain Miller visited the shop a couple of other times, buying a scarf or two. Each time he went to the shop this man would point at the cross on the uniform and back at himself. The pair would visit briefly and then part ways.

Chapter 36

Christmas

Hordes of packages began to arrive for the troops for Christmas. One of the best and most useful presents Chaplain Miller received was a knitted cap and a pair of thick wool socks. He remembered when he opened them nearly shouting, "Merry Christmas to ME!!" They came in a care package from some of the folks from Camp Robinson in Arkansas and immediately became part of his nightly bedtime garb. It got extremely cold in his hut at night and when he got ready for bed, he put on those thick socks, covered his head with the knitted cap, and zipped himself up in his sleeping bag, praying before he went to sleep that the Taliban would leave them alone that night because he was now warm and toasty.

Chaplain Miller had his mother send him a Santa suit, which quickly earned him the nickname "Combat Claus." He also had her send an elf suit and he managed to get Chase to wear it, but not without some serious convincing. So, imagine this guy much shorter than Chaplain Miller in an elf costume and fully kitted up for battle, wearing his Kevlar, the different belts, and his M4 strapped across his chest. It was a sight to behold.

About a week before Christmas, they began making the rounds to the different outposts wearing these suits. Chaplain Miller would put on his uniform and then put his Santa suit over the top of it. He would pull the pants on, fasten the suspenders, slip on the jacket, and put his Kevlar vest on over the jacket. The beard would hang over the top of the vest and when he took his Kevlar helmet off, he had on the Santa hat. These two were quite the sight as they traveled across the country, delivering coffee, mail, and presents. They would show up on the flight line ready to leave and the air crews would shake their heads as they boarded the various aircraft. Some really loved it. They wanted to pause for pictures, and they thanked them for brightening the day.

When they landed at an out station, Chase would have a Santa sack filled with various types of coffee and would begin handing these out to the troops. The troops Ate. It. Up. They would visit with the guys for a bit, pray with them, and then move on.

The best trip was on Christmas Eve. The commander was going on this trip and he was totally unaware that Chaplain Miller and Chase had been dressing like this. He walked up to the vehicle that would take them to the flight line and saw Santa sitting in the vehicle. He walked completely around the vehicle until he noticed Chase sitting in the back. He looked at Chaplain Miller and said, "You have a freaking elf, too?" Everybody rolled with laughter.

CSM Stain was dealing with a soldier issue and was running late. The commander said they had to leave without him. They got to the C-130 and the crew wanted a picture with Santa and his elf. They posed for all of the pictures and then boarded the aircraft. Suddenly they heard CSM Stain shout at the loadmaster, "Is there a Santa aboard this flight?" The loadmaster walked to the back of the plane and yelled back, "Yes, and he's got an elf with him!" CSM Stain responded, "That's great because it means I have the right flight," and he boarded the aircraft. He had been running down the flight line asking each plane if there was a Santa on board. This flight went to Kunduz. There was a group of Afghans and a group of Germans guarding the flight line and it was hard to discern who had the biggest smiles on their faces. Chaplain Miller is sure there are photos floating around Germany somewhere of him and Chase because these guys were taking tons of photos with their phones.

This was how he and Chase prepared the troops for Christmas. They knew Christmas was coming, and here they were stuck thousands of miles away from home and probably bored to tears at this point, as well as being freezing cold. He also knew this was a time when depression would rear its ugly head and he wanted to make sure these guys had something to laugh and smile about. It was amazing what the sight of two guys dressed like Santa and an elf in their battle gear and bringing coffee will do for a soldier's morale. The visits weren't very long, but they were extremely effective.

After visiting and passing out coffee and presents to the troops, Chaplain Miller would pray with them and head off to another section. He knew he had done the work God had put him there to do. No, he didn't beat them about the head and shoulders with his Bible or preach some sort of evangelical sermon. He followed the command of Jesus that says, "Love your neighbor as yourself." He knew what their mental state was because he was experiencing the same thing and did everything in his power to change their perspective.

The next day was Christmas, and Chaplain Miller did what so many others were doing. He opened his presents from his family and called home to talk to as many of them as he could. This scene was repeated all across the bases and outposts around Afghanistan and, frankly, on every military base and outpost around the world. It is a scene that gets repeated year after year since we always have troops stationed away from home. And true to their word, that evening the Taliban launched some rockets at Bagram. The sirens warning of incoming rockets began to sound. Both Chaplain Miller and Chase moved begrudgingly to a bunker. As they sat in the bunker, they both expressed out loud the thought that this was not where they wanted to spend any further Christmases. After hearing the rockets impact elsewhere, they stretched and went back to their calls and emails.

Chapter 37

January

Fighting season had slowed considerably. That doesn't mean all combat had ceased because there is always something happening somewhere, but as far as major operations were concerned, it had ground to a halt. While many elements of Jeremy's group guarded towers and checkpoints, others still went on the occasional patrols. Still, if the intelligence pointed the right direction, their task force stood ready for anything from raids to training local Afghans.

January was the halfway point for the deployment for this specific group. Troops began to, whether they realized it or not, look forward to getting the next six months over with because it meant they would be going home. With fighting having slowed down, things became sort of mundane. Troops were doing their jobs, but there was not as much to do outside of those duties. It actually became a little boring at times. This is another instance where the term "embracing the suck" comes into play. It's not because things are hard; the missions and shift work were long, but at least most troops found a sense of accomplishment in that work. The suck comes because now soldiers have time to think about things. They begin to ask the question, "Why am I here?" and their minds begin to wander to things other than the mission in Afghanistan.

Chaplain Miller found his job evolving during this time. The services he was preaching continued to be well attended, but his time spent counseling increased dramatically.

Every little problem, every little rub, every little issue now became a major moment in soldiers' lives. If two soldiers were sitting in a tower and one of them has a little tic of some sort, at the end of their shift they wanted to strangle each other. If several soldiers were on a convoy and one member of the group kept telling the same lies over and over, then the group was near ready to toss them out upon arrival back at their outstation. Typically, when a problem occurred, the soldier would go to their platoon leader who sent them to the first sergeant who in turn sent them to the chaplain if a strong scolding to stop acting like children didn't fix the issue. Items like pay and promotion issues started showing up, and those were the simple things with which to deal. However, there were other issues that were much more serious that began to rear their ugly heads. We will start with pornography.

With today's technology, getting to a porn site is simply a click away. If a soldier has access to WiFi, they have access to porn. It is just that simple. Add the element of boredom and not having any other place to go or anything else to do, and it's really simple to take that next step. We really can't elaborate on the kind of stuff they could find and watch but it was, and is, a serious problem among troops all over the world, especially in a combat area. Chaplain Miller constantly stressed to his soldiers to "not go down the road of pornography because if you do, over the course of the next year you will train your body and mind to think differently and react differently. And when you go home, your mind is going to expect your spouse or girlfriend or boyfriend to act and perform like the pornography you have been watching for a year." When it was warmer and a lot more active, the troops didn't have time to sit and watch porn when off duty. But now that it was colder, the activities became a lot more confined to their living areas. It was too cold to get out and exercise, play basketball, etc. They only went out to eat, man their duty stations, and maybe exercise at the closest indoor workout point, but other than that, they spent a considerable amount of time in their living quarters.

The second major issue for counseling revolved around infidelity, not from the troops but from their spouses. There was one example that stood out to Chaplain Miller.

A soldier came into his office and said, "I think my wife has been cheating on me." "This is a pretty serious thing. How are you going to handle it?" "Well, I don't know because she's cheating on me with another guy here in the unit." Chaplain Miller's mind kind of fogged over because he was instantly trying to figure out how this was happening. So he asked, "She's cheating on you with another guy that is stationed here in Afghanistan with us?" "Yes, chaplain." So, Chaplain Miller said, "Can you say that one more time because it's not making sense to me." The soldier held out a stack of paper and said "Here are the texts messages from her online chat that she has been sending back and forth to this guy. They are preparing for whenever we get back home and what they are going to do." Now it made sense.

They talked for a little bit and Chaplain Miller addressed things about what a Biblical husband looked like. Then he asked, "Do you want to forgive her?" The soldier replied he did and he wanted to make his marriage work. Chaplain Miller realized he had two guys who were bunking beside one another for the next six months with weapons at their disposal. Needless to say, the consequences could be deadly.

Chaplain Miller went Old Army (old school, for everybody else). He put the first soldier in his office and called the first sergeant to have the other soldier report to the chapel and, logically, not bring a weapon. The other soldier showed up and Chaplain Miller moved him into a meeting area and locked the door behind him. The soldier asked, "Am I in trouble? They took my weapon from me and told me to report to you." "No, you aren't in trouble. Just have a seat, and by the way, do you have your knife on you?" The soldier got a quizzical look on his face and replied to the negative, which eased Chaplain Miller's tension level a little bit.

He then brought in the aggrieved soldier and had him sit across from the other one. Both of them looked at him with looks that said, "What's going on?" because neither knew the other was going to be there. Chaplain Miller sat between them and said, "Here's the deal, gentlemen. You guys have no choice but to get along for the next six months." Looking at the first soldier, he said, "You have a wife," and then turning to the other soldier, "You have what you think is your girlfriend, and they are both the same woman. So here's what I'm going to do. I'm going into my office and send a few emails and you two are going to stay in this room and work this out."

Needless to say, both soldiers got very tense very quickly. The one who had been sending the texts and emails was looking for a way out of the room, which happened to have no windows and only one door. He looked at Chaplain Miller and begged, "Please, don't leave me. Please!!" Chaplain Miller asked him, "So, what do you think is going to happen?" He said, "No, Chaplain, I would like to talk about this with you here." "OK, I'll stay for a few minutes but I am eventually going to leave the two of you in this room together. That door will remain locked and you two will talk about this." "Chaplain, there ain't no talking about it." He was really fidgeting at this point. "Okay, do you want to just go ahead and say what you've been doing?" At this point, he spilled the beans. The other soldier had all of the evidence anyway and all of this was verified by the soldier who had been sending the communiques. This guy laid out everything to clean the slate.

When he finished, he said, "I did all of this. I'm such an idiot. Can I go back to work now?" But Chaplain Miller said, "No, I've got some emails to check on and you two need to talk." The soldier was emphatic, "I don't want to talk, and I don't want to be in the room alone with this guy." Then Chaplain Miller looked at them both and said, "Now listen, do not do anything that is going to get either of you in trouble. Neither of you will leave this room until I say so, but both of you need to go back to your jobs after this is finished. Now, I'm going to step back into my office and I'll see you two in a little bit." With that, Chaplain Miller left and went to his office.

For a little bit, there was this eerie quiet, and Chaplain Miller was afraid they were choking each other. But after a little moment he heard one of the soldier's crying and saying, "I'm sorry. I'm so sorry." And then he heard the other soldier speak many of the same lessons of forgiveness Jeremy had shared earlier in counseling. The soldier said, "I forgive you. You are my brother, and I am going to choose to forgive you. I'm not going to bring it up again to you and we are not going to punish you. I'm not going to harbor this over you. I'm not going to do any of that. But you have to promise me this is going to stop. There will be no more communication with my wife." The other soldier replied with what sounded like, "I still owe you. Please punch me in the face or something." Secretly, Jeremy hoped the soldier's plea was not accepted so he wouldn't have to explain things later to the CSM. "No, I forgave you, and that is what that means." There are not a lot of men that could and/or would do that. Most would probably have hit him in the face, but this guy chose not to. Chaplain Miller said it brought home to him that people were actually listening to his counsel and were acting in a way that Jesus would want us to act.

One of them knocked on the door, and when Chaplain Miller went back into the room, both of them were red-faced and teary-eyed. He said to the soldier sending the messages, "Now look, you flirted with and tried to entice his wife into leaving him and he should just whip the dog out of you and I should be OK with that. However, he chose to forgive you, and I am definitely OK with that. Now, here's how you walk in forgiveness." And he walked him through Romans 1. He showed them both how this applies to all of us. "It shows us that we are all sinful and depravity runs in our veins." Then, looking at the first soldier, he asked, "Do you have a porn issue in your life?" When the guy couldn't answer and looked away, Chaplain Miller said, "I'm going to take that as a yes. Due to this porn issue, you've probably been neglecting your wife, not calling her as often as you should because you are locked in your room trying to gratify yourself with something wrong and different. That needs to come to a halt, and if you are truly going to forgive her, don't give your wife a reason to seek after another man." Looking at the second soldier, he said, "Be thankful you haven't been punched in the face. You walk in forgiveness by never going down this path again. A dog can return to its own vomit and a fool is going to return to his folly. Are you going to be a dog and a fool?" He answered, "I'm not a fool." "Let me rephrase that," Chaplain Miller said. "In Scripture, a fool means an idiot returning to its mistakes. You just made an idiotic mistake by texting and emailing another man's wife. Don't ever do that again. Do you understand me?! Your leadership does not need to know about this, and we are going to keep this between us."

By the end of the deployment, Chaplain Miller knew they had shared it with others because word spread what had happened. Some called it the "coming to Jesus counseling with the chaplain." Regardless, others were amazed they had not beaten the mess out of each other. This was an instance where a chaplain could work outside the chain of command and rectify a situation that could have become detrimental to the entire unit. Their commander was a young lieutenant and would not have had a clue on how to deal with this situation. A chaplain can insert himself into this kind of issue, keeping this from becoming a command problem, and actually save the careers of a couple of good soldiers who otherwise would have likely been put out of the service. This was indicative of the types of counseling Chaplain Miller performed for the first couple of months in the New Year.

It is these instances that bring up the difference between a chaplain and a pastor. While a pastor is familiar with their congregation, a chaplain literally lives and walks among their soldiers daily. The situations they handle are much different than a normal pastor has to deal with. Not saying that a pastor doesn't deal with infidelity or pornography issues, but the settings are much different. Also, church members are very reluctant to go to their pastor with these things but soldiers readily either go or are sent by others higher in their chain to the chaplain, and these issues are discussed on a regular basis.

Since the end of November, Chaplain Miller had been the only chaplain within the entire task force to which his detachment was assigned. This meant an increased workload. Also, even though the fighting had slowed, as mentioned earlier, there were still occasional engagements. When those happened, Chaplain Miller and Chase would hop on a transport and head to these locales to do whatever needed to be done, of course, depending on the weather. When there were wounded casualties, he would have to provide counseling for those involved. Adding to this, there were a number of emergency surgeries, e.g., gall bladder surgeries or broken bones. He would have to be there for those and deal with the soldiers and their families as well.

There was a group of Army Rangers close to Kandahar. Their chaplain, still in the States, called Chaplain Miller and asked him to do services for his troops. When he asked why, their chaplain replied, "Because they trust you. They've seen you when they are preparing for missions and when they get back. Many of them have heard you before. You told them it was OK to hurt but not OK to stay that way, or something." Chaplain Miller got a laugh. This was a theme he carried throughout his time in-country. There were a lot of young people seeing more than their share of bad things and not handling it well. He was consistently relating this mantra: It was OK to hurt or feel badly but it was not OK to stay that way. It's OK to grieve but when done, be prepared to move forward. What he was trying to do was get them to understand there were bad things happening and it was OK to know what happened but not to let that one thing they had no control over define them for the rest of their lives. PTSD is a real thing, but he was trying very hard to get them to understand it but continue on with their lives. Rather than succumb to a label of a "disorder," he wanted them to focus on the growth they could have after the trauma.

Chapter 38

February

February continued at a slow pace. There was one small incident of shelling which didn't injure anyone or affect anything the troops were doing. Any mortar or rocket fire was merely an annoyance now. Soldiers would stop, kneel against a solid surface, and listen to see if the rounds were going to land far away or not.

Soldiers had more important matters at hand. The Super Bowl was going to be played and, as has been usual for what seemed the past twenty years, the Patriots were one of the teams in the game. As stated earlier, the Rhode Island contingent of the unit were huge Patriot fans. In one location, most of the troops were from the Boston area. Their TOC and quarters looked like a Patriots memorabilia store. Everywhere you looked, there was a Patriots flag or banner, Patriots gear on desks, and if truth be known, Patriots shirts were probably being worn under their uniforms. Chase was in football heaven. He said, "All you can get in Arkansas is Dallas Cowboys stuff and I hate them." Every time he went around that unit, they were giving him something, e.g., hats and scarves, and he said it was like getting to have Christmas all over again. He spent as much time as possible with those guys.

From Christmas Day to the middle of February, there was very little Taliban activity directed at Bagram. Fighting the mundane became a daily fight. Examples of mundane were reinforcing barricades, vehicle maintenance, and catching up on reports. The troops knew it wouldn't be long before they were headed home. Leaders were working hard to keep troops focused while 1) they are not getting rocket attacks on a daily basis and doing routine tasks and 2) their knowing it is just a very few months before heading back to their loved ones. The calls from first sergeants and company commanders became more frequent. With weather like it was, it was difficult to travel to some of these outposts, so consultations were done on the phone with the troops or through video chat where it was available.

Chaplain Miller was not immune to the dealing with the mundane. He wanted to be out with the troops, rolling with them on missions, and working to keep their morale up. Sitting in the chapel or his office all day, every day, was not his idea of being able to help the troops. He was itching to go and simply could not. Patience is not an easy thing for a deployed soldier to practice or exhibit. However, 1 Samuel says, "lack of patience can cause you to miss blessings." Being patient is a vital part of trusting God, as our life experiences and expectations are not always what we would prefer.

One thing Chaplain Miller did to help break up the mundane a little occurred about two weeks before Valentine's Day. He procured a bunch of Valentine's cards and began walking around to the various offices of his troops. He told them to look through the cards, pick one, and write something to their spouse or girlfriend/boyfriend. He told them he would be back at the end of their shift to pick them up, and he put every one of those cards in the mail for them. Towards the end of the month, soldiers began seeking him out to tell him thanks for bringing the cards and making sure their loved ones got something from them. It was something to help the troops understand where they were, what time of the year it was, and to help their minds think of something different.

The leadership also did something to help quite a bit, based upon the recommendation of Chaplain Miller and others, including Chase. They changed the routine of the soldiers. They changed the shift work. Those who had been working nights were moved to days and vice versa. They changed up the assigned tasks. It caused a complete mindset change because each shift had different things to deal with, and getting a new shift and new situations meant they had to reengage their minds and thought processes.

Chase was also one of those who would do different things. Chaplain Miller remembered coming through a checkpoint and when he looked up to show his ID, Chase was the one doing the ID check. He said the troop that normally handled that duty needed a break and Chase gave him one. Chaplain Miller was not aware that he had been doing this for a couple of months. He told Chaplain Miller that when he didn't have something to do, it was a way to help others and to also give himself something different on which to focus.

Chaplain Miller was still preaching and counseling as much as he always did, but his travel was curtailed significantly because winter was still going strong. He hated not being able to travel. As normally happens, when we get in a rut, things happen to us, and it was no different on this deployment. One day Chaplain Miller was coming out of the chapel and slipped on some ice. He went down in a heap and after checking himself to see if he was all right, he discovered, when trying to stand, he had injured his foot. He went to the medical clinic to get it checked and the X-rays revealed he had several small fractures in his foot. The doctor told him he was going to have to be in a walking boot for the next three months. In no uncertain terms he stated that wasn't going to work. With a walking boot, he wouldn't be able to get out and visit his troops, work out at the gym, or basically do any of the stuff he thought was most important. After a bit of wrangling with the doctor, they agreed to a week and a half in the walking boot. Men truly understand this and women just roll their eyes and shake their heads.

February was cold, wet, and Chaplain Miller was deployed in a walking boot. He tried to wear a plastic bag of some sort to keep the water off of his foot and it slowed him down in a way that otherwise would not happen. He was not happy with this situation because it was not his nature to sit around, and he was having to do that because of his foot. He became sulky and hard to live with. But God had a use for this injury. Chaplain Miller had been representing a limitless and tireless God, and he was beginning to think of himself as limitless and tireless.

This is the wonder of God. Even though he didn't realize it, God knew Jeremy needed a rest, a chance to recharge for the home stretch. God knew what was coming and knew Chaplain Miller needed to be at his best. It is kind of like God rejuvenating Elijah by giving warm bread and water upon his awakening and having an angel touch him twice. It takes great faith to admit we are tired and need a break. It goes against our nature to rest, because to rest, we have to trust that others are taking care of things. We have to trust that God is in control and we are only doing His work as He lays it out before us.

Making Chaplain Miller rest wasn't the only thing God did for him. All of his counseling dropped off to nothing during this period, and all he had to do was preach. This allowed him to get into the Scripture more and delve deeper into what God had in mind for him to preach. It allowed him a time to seriously reconnect to God and what God had in mind for him. One of the pieces of Scripture God revealed to him was in Romans 1, particularly verses 16 and 17 (NIV): ***"For I am not ashamed of the Gospel because it is the power of God that brings salvation to everyone who believes; first to the Jew and then to the Gentile. For in the Gospel the righteousness of God is revealed – a righteousness that is by faith from first to last as it is written: 'The righteous will live by faith.'"***

After reading this particular piece of Scripture, Chaplain Miller had a conversation with God. There is a lesson here. First, a conversation is a two-way discussion, and second, to have a conversation, you have to listen to the other person. Chaplain Miller, though still upset about his predicament, was faithful enough to listen to what God was saying. If the conversation could be put into words it likely went something like the conversation God had with Job when He told Job he was going to listen to God. So imagine this scene:

Chaplain Miller: Lord, I can't do what I need to do while I'm sitting here with this boot on my foot. (Probably said more out of anger or prideful arrogance.)
God: To slow you down, I allowed this small accident to happen.
Chaplain Miller: But that means I'm stuck here and can't do the things I want to do.
God: Your faith in representing Me is why I had you here in the first place. It is not about you.
Chaplain Miller: But..... (You could insert a laundry list of things he was ticked for missing out on, but God stopped him mid-sentence.)
God: You aren't in this place because you are some physical guy or great chaplain. You are here to represent Me.
Chaplain Miller: So, you think I may have lost a little focus?
God: Bingo, genius!

Chaplain Miller: So, this is going to take a while to get over?
God: A little, but through this time I can bring your focus back to where it needs to be.

Chaplain Miller continued his study in Romans. Romans 1 is a scathing indictment of what a godless society looks like and also shows the only path from that society. Continuing this study in Romans brought home to Chaplain Miller that he had become so busy preaching and counseling that he was beginning to fail to live it in other facets of his life. He did begin to understand this rest was important and allowed him to live his life closer to Christ than he had in the past couple of months. Rest was indeed important, and it was sad that God had to allow him an injury to reveal this to him.

He took this message to heart and began to not only try to live it out but also began to preach this message to everybody. It brought home to Chaplain Miller that he needed to get laser focused on what God really wanted him to do, and that it was particularly important during this time when they were getting only one or two inbound rockets a month. It was important that he preach about taking care of themselves and their business, even during this mundane and boring time, because there would come a day when it would not be mundane and boring.

The primary subject he had to deal with during this time was the question, "What am I doing here? Why am I in this place?" When soldiers are away from home and not really involved directly with the enemy for long periods of time, their minds begin to wander to other things. The mind can be a terrible thing if allowed to wander unchecked. Soldiers of all ranks began to think of home and what they were missing. They began to think about wives and girlfriends and what they were doing. Thoughts of children's birthdays and other milestones being missed were pushed to the forefront of their thinking. There are a hundred other things a soldier can think about when they have the time to think. Chaplain Miller found these situations and thoughts permeated his counseling sessions.

When you take those thoughts and add to it bad news from home, it can be a real morale breaker. It seemed like January and February were the peak times when bad news hit. Numerous soldiers received word from their spouses or girlfriends that they wouldn't be there when the soldier got home. There were also a significant number of babies born and soldiers could not be there for the births. They would get this kind of news and then have to pull watch in a tower or sit in a long, boring convoy traveling to another post while these situations were at the forefront of their minds. This nearly two-week period of time when Chaplain Miller was down and heavily immersed in the Scripture allowed him a new life view and perspective when he was finally able to get out and start dealing with the troops again.

When dealing with a soldier experiencing these issues and thoughts, he came across a couple of questions to ask. The first one was, "What thing taken from you would cause you to walk less closely to God?" The answers varied among kids, wives, jobs, and the uniform. The uniform answer wasn't about that particular uniform. The soldier was talking about having a job.

As guardsmen, when they got back home, they would be going off of active duty and many of them didn't know what they were going to do because they would not have a job waiting when they got home. Soldiers were sitting in a gun truck all day long and the thought in the front of their mind was about what they were going to do when they got back home. Not looking out for enemy attackers, not looking out for IEDs, not keeping their minds on what might be coming from around the next corner. This break enabled Chaplain Miller to be able to focus on what the troops were truly dealing with and counsel and help in a much more efficient manner.

The second question was, "When was the last time you took a rest?" Chaplain Miller now understood what this had meant to him, and he wanted to pass this along to others. Of course, their initial answers were like his had been. "Oh, I take a break every now and then." This would give him the opportunity to discuss the difference between a break and a rest and to discuss the ramifications of both. He had one officer ask him why a rest was so important. Chaplain Miller answered because first, by trying to do it all himself, he was not being an effective leader and mentor to his troops. And the second reason was, most importantly, that eventually he would crash and be taken out altogether. Nobody would be able to do the job if their leader crashed.

Chaplain Miller began to live what he was preaching to everybody else. His routine became different, allowing for periods of true rest, periods when he was not doing anything. Instead of using his extra time to go to the gym and work out some more or find someone to hang out with, he would head to a place of peace and quiet and recharge. It became noticeable to those around him what he was doing, and others began to follow suit.

However, on 18 February 2012, things got back to normal. There were, as Chaplain Miller described it, "a few rockets fired at Bagram." He said you could see the dullness leave the eyes of the troops, and suddenly there was a new spring in their step. It was like, "YES, it's back and we are back." It was time to get back to what they were sent there to do: contact and defeat the enemy. It was not an attack like at Jalalabad in December. However, those few warning shots on that day in February changed the entire demeanor of the troops.

Chapter 39
March

March brought to home that there were only three months left before heading back to the States. It was brought into the limelight when the advance party from the unit replacing them arrived. The advance party's job was to arrive, assess the situation they were coming into, and go back to their unit to do their final preparations for deployment. Chaplain Miller and Chase were hoping they would be getting good replacements. The Executive Officer with the advance party handed Chaplain Miller a list of questions from their chaplain who would be replacing him, and he was relieved to see, based upon the questions he had been given, he was being replaced by a good and caring chaplain.

It also brought another item for Chaplain Miller to be aware of. Troops knew when it was time to rotate out, and they became excited about that upcoming period. Unfortunately, in some cases, too much so. When Chaplain Miller would encounter this problem, his primary message was to stay on point. The message went something like this. "Yes, I know in a few short months we will be leaving. However, we have a job to finish here and we need to stay focused on that job and mission. You can think about going home at a later time, but in order to get to that point, we have to do our jobs and do them well." He had to reinforce this message at every station he visited.

Kandahar had a place that was interesting to the troops stationed there, and Chaplain Miller tried to visit it on every trip that took him close enough. It was called the Boardwalk, which actually might be called an International Bazaar. It was a place where troops could get away from things and relax for a bit. In the middle was a large recreational field where soccer and other sports were played. It was surrounded by a covered boardwalk, allowing shoppers to walk above the ground and be relatively protected from the elements. They could buy purses, bags, rugs, DVDs (most, if not all, pirated), and all manner of items locally produced.

He went into one of the stores that sold books and notebooks. They were all leather bound with nice, pressed paper. He found a journal that was pocket size and bound together by string. The cover had the word Afghanistan pressed into the leather, nicely done, with a green stone pressed into the center of the leather. It caught his attention and he bought it for his son.

What spurred this purchase was talking to his troops about focusing toward the end and not making a stupid mistake. That last conversation had been interrupted with incoming rockets at their outstation a few days earlier. In the end, the shots weren't that close but caused Chaplain Miller to start thinking about his young son. The thoughts that ran through his head were along the lines of, "What if something happens to me. He's a young child. Who's going to help him grow up and teach him the right things to do?" So, with those thoughts in his mind, he purchased this journal and began writing in it. He took it everywhere, and as thoughts came into his mind, he would jot them down. These would be the lessons he would want his son to know if he wasn't around. He used things from his past, things he learned from others, experiences he could pass on (both good and bad), and life's lessons that had been experienced. Basically, it was like the book of Proverbs but specifically for his son. There were about thirty lessons, or proverbs, in the finished product.

Chapter 40

Attack

Three days in March stood out to Chaplain Miller: the 11th, 20th, and 27th. The reason those three days stuck out to him was Bagram was attacked on all three of those days. The irony was that he had returned to Bagram the day before from visiting outposts on each of those days. In between times, while Chaplain Miller was traveling, nothing was happening at Bagram in regards to attacks. It was only after he got back from each trip that things picked up. Chaplain Miller remembered the chaplain he replaced complaining about nothing ever happening there. This was definitely not the case with him.

On 11 March, Chaplain Miller and one of the section sergeants, SFC Jeu, rode across the base to check the tents that would be used when the troops began returning from the various outstations and preparing to return home. This would be happening in less than two months. While they were checking the tents, they discovered a coffee place called Green Bean and went there when they finished their check on the tents. While they were standing in line, the sirens went off signaling some sort of incoming fire. They scanned the area and discovered there were no bunkers around. Add to that they were not "kitted up," meaning they were not wearing their Kevlar vests or helmets. That doesn't mean they didn't have them. They were in their Humvee, just not on their bodies. Protection for them at this time was minimal. They got behind a concrete barrier and lay head to head for protection until they knew what was going on. One thing they made sure they had, though, was their cups of coffee. Once the first rockets went overhead, SFC Jeu jumped up and said, "Let's go." Chaplain Miller asked if he was sure, to which Jeu said "They are going over our heads, which means we won't get hit. Let's get out of here." Throughout this rocket attack, they didn't spill a single drop of coffee.

20 March was a different sort of event. Chaplain Miller and Chase returned to Bagram the morning of the 20th and got rid of all of their gear from traveling. Chaplain Miller went to the TOC and reminded everybody he would be doing Bible study that evening at the chapel lounge. That afternoon, he was sitting in the chapel annex preparing for the Bible study and praying. The chapel annex had a nice atmosphere, compared to some other places on Bagram. Troops could relax, and the place had a sort of lounge feel. There was always the aroma of popcorn and coffee. This might seem odd to some people, but for GIs, it makes perfect sense.

Chaplain Miller had put the closed sign on the front door because he did not want to be disturbed, and he got comfortable. There was also silence, the type of silence that can be hard to find on an active and combat ready base. In fact, it was so quiet and comfortable, he was fighting the urge to sleep. He started talking to God about the upcoming Bible study and thoughts of how these studies were starting to positively impact some lives. As he was praying, he started seeing the faces of these people whose lives were being changed by these studies. He realized he was becoming extremely connected with these soldiers in a way so unlike a traditional pastor does with their congregation. Seeing these people and working so closely with them day in and day out had built relationships that can only be built in that type of situation.

He was studying the book of Acts, which deals strongly with how the church was built, and following the commandment that Jesus had made when He said, "Go!" Then Jeremy asked this question of God. "What are You laying on my heart?" He quit praying, sat up, and realized there was no one in the building. Suddenly, Chaplain Miller begin to feel as if God was impressing on him he needed to get out of the chapel. The impression he got was God telling him He didn't put him in this situation to minister within the four walls of the chapel. Chaplain Miller, not arguing with God but just looking for answers, thought, "Lord, we have nice couches here and a much better setting. What has changed now?" But the impression to be somewhere else became even stronger. Chaplain Miller grabbed his stuff and walked over to the TOC, leaving the closed sign on the door of the annex. Looking back, he believed God moved him from this place.

If you remember the television show Cheers, there was always a scene when Norman entered the bar and everyone yelled, "Norm!" greeting him as he arrived. This was the type of welcome Chaplain Miller got when he arrived at the TOC. "Chaplain," and, "Chaplain's here" were the shouts that came from the troops. They asked if he was doing Bible study there and when he affirmed, they passed the word around to everybody. About five minutes later he started into a side room next to the TOC and realized he had more people there at that time than he had ever had before. As he looked around the room, he saw the same faces he was seeing earlier as he prayed and realized it was one of the best attended studies since they had arrived. Obviously, God was putting these soldiers in his heart and mind for a reason.

Work kind of stopped, and he began the study. He asked if there were any questions and he remembered they asked some really good, deep, thought provoking questions that evening. He remembered having this huge amount of joy as he finished because he knew these people were becoming impacted by what they were hearing and reading. He ended the study and was on a high note as he left the room next to the TOC. But as they began to step outside, a rocket zoomed just overhead and exploded near their area. No sirens, no warning. All he could hear was the sudden screaming of the rocket as it shook the room and the gravel beneath his feet. The impact caused everybody in the TOC or outside to dive for cover. The whole building shook as the rocket hit a building right next to the chapel and chapel annex, exploding and spraying everything with shrapnel. Major Wes Dumas dove to the ground next to Chaplain Miller. Wes had been involved in the firefight that killed Saddam Hussein's sons in Iraq. He looked at Chaplain Miller and said, "That's the closest one's ever got." There was nothing protecting the chapel and the chapel lounge, and it was sprayed heavily by the shrapnel, especially Chaplain Miller's office but more especially the lounge. Keep in mind the chapel, the office, and the lounge were all constructed of plywood, meaning there was nothing to stop the shrapnel as it moved from the blast point.

Chaplain Miller remembered the looks and reactions of individuals who had been under mortar and rocket attack. He remembered his ears ringing, not the type of ringing where one can't hear anything, but a ringing that made it difficult to hear. When they heard this rocket, they were ducking for cover and grabbing everybody who was slow to react.

There is a difference in sounds between a rocket going over like the one on the 11th and one that is falling on your location. There were also people who were just leaving the TOC when the rocket struck. They saw it go overhead and impact and, fortunately, none of them were hurt. They did have to check out one of the soldiers in the guard post close to the chapel for a concussion, but this was just precautionary. Chaplain Miller and some others headed to the chapel to assess the damage, because they knew it had been hit.

Moments earlier, Chaplain Miller had been filled with joy from what he had experienced through God. Now, he was full of adrenalin and trying to figure out what to do with this sudden rush of emotion. His first thought was someone just tried to kill me and take away all of the joy I was experiencing. He was so mad about it and, in all honesty, did not know exactly what to do at that moment. Being a former artillery officer, he knew within moments there would be outgoing rounds trying to kill that faceless enemy that had tried to kill him. He did get some satisfaction knowing what was about to happen. He didn't exactly follow the Scripture that says pray for your enemies except to pray they were dumb enough to hang around for the artillery that was coming their way. This was a reminder he was a soldier as well as a chaplain.

Upon arrival at the chapel, the reality of the situation hit Chaplain Miller. He walked to the door, flipped the closed sign over to the Open side, unlocked the door, and walked in. He could immediately see the holes the shrapnel made as it traveled through the chapel and the annex where he had been sitting. If he had been conducting that Bible study at the lounge like he was preparing to do, numbers of soldiers would have been injured or killed. The shrapnel traveled through exactly where they would have been sitting or standing, both inside and outside of the chapel lounge. The thought came through his mind of the Scripture about standing before God, and he knew God had told him it wasn't his time. On reflection, Chaplain Miller realized God had moved him and the study from a place of danger to a place of safety. "Be still and know that I am God" truly applied at that moment. If Chaplain Miller had not listened and reacted to what God was telling him to do, lives would have been lost or changed forever.

The anger he had been experiencing moments earlier moved now to the realization of how much smaller he was than God. God was moving him, in spirit, in flesh, and in faith. All in a period of less than two minutes. He became truly in awe of God, and all he could think was, "Thank you, God, for taking care of me." This was his "aha" moment in his relationship with God. He experienced three emotions within a period of less than five minutes: joy from the tremendous Bible study, anger from almost being killed by a faceless enemy, and awe from suddenly and truly realizing the power of God.

It also changed a little of the paradigm under which Chaplain Miller had been operating. Recall how the Bible study in the TOC had been attended. This drove home the impression God was putting on him. Not only did God keep them safe by moving them from the chapel, but also more people attended and heard the Word of God. Bible studies began to happen where the soldiers were. "Go forth" is a common theme in the Bible, and Chaplain Miller took it to heart. One thing he did get out of this attack was the Combat Action Badge, which he will wear forever. To him, it means the enemy tried to kill him and was not successful.

He also picked up something that helped him counsel with soldiers during the rest of the deployment. He had two reactions after looking at the results of the rocket attack. One was thankfulness to God for protecting him. The other was an overwhelming anger at the guys who had tried to kill him. This was a reaction he had encountered as he traveled from post to post but one he didn't fully understand until now. He could relate to what soldiers were experiencing after engagements with the Taliban, and this helped his counsel become more effective.

The next day, Chaplain Miller and Chase left again for outpost visits and returned on the morning of the 27th. Chaplain Miller got rid of his travel gear and headed for the TOC. When he got there, he asked if anything had happened while he was gone and was told that it had been quiet all week. It wasn't thirty seconds later that another rocket attack started. This one was a lot like the one on the 11th with most impacts far enough away, but he began to feel like he was becoming bad luck. One soldier in the TOC even jokingly stated that very thought. It was good they could joke after such an eventful time.

Chapter 41

Funeral

Before deploying, Chaplain Miller wrote out his own funeral. Everybody was told to write a Last Will and Testament before they deployed. Chaplain Miller took it a step further and wrote his entire funeral. He only told one person, a good friend of his. He told him it was in a special location in his house and he was the only one who knew it was there. In the letter, he outlined everything he wanted done at his funeral. He didn't want Beth to know about it because she had enough to do right now and did not want her worrying about that aspect of the deployment. Because of his experience doing death notifications, he had experienced too many times a wife saying, "I don't know what to do or where to go." He did not want Beth to be in that situation.

He laid out what songs he wanted sung, who he wanted to preach the funeral, and what he wanted them to talk about. They could use whatever verses they wanted to use, but he wanted a specific topic discussed. He laid out an order of who would talk and when they would talk. He laid out where he wanted the funeral to be held and the format of the service. The last thing he wanted was a soulful dirge. He wanted a celebration of his life, because it was a life to be celebrated.

In addition to the funeral arrangements, he also wrote a paragraph to each of his children. These were special and extremely personal messages. It was emotional for him to write and difficult to do prior to deploying, but it was something he wanted to have done before he left. The hardest one to write was to the child that was yet to be born.

Since he didn't become a casualty while deployed, shortly after coming back home, he showed this letter to Beth. She was totally surprised by it and Jeremy told her they would keep it and use it down the road. In a brief side note, that letter now has significant updates. People who were to have spoken have either died or moved away. Circumstances have changed, and other than the messages to his children, these instructions would be hard to adhere to. But they were extremely applicable for his deployment, and he was glad he had taken the time and effort to write them out. These instructions are another example of how an individual should be prepared to go at any time. ***"Why, you do not even know what will happen tomorrow. What is your life? You are a mist that appears for a little while and then vanishes."*** James 4:14 (NIV). We are not promised tomorrow, and once that final breath is taken, it is finished on this earth. If you do not have a relationship with God at that time, it will be too late.

Chapter 42

Red Cross

When emergencies happen at home, one of the primary ways of getting notification to the deployed service member is through the American Red Cross. The family member notifies the Red Cross and the Red Cross, in turn, notifies the unit of the family emergency. Prior to the days of cell phones and social media, this was the way service members were notified of home emergencies. The American Red Cross does not authorize emergency leave for members of the United States military. The Red Cross's role is to independently verify the emergency, enabling the service member's commander to make an educated decision regarding emergency leave, and then to provide transportation assistance and/or financial assistance if needed.

For some reason, Red Cross messages always arrived in waves. There were the good kind, which primarily consisted of births. These notifications can be delivered by the commander, first sergeant, or any of the leadership. In Chaplain Miller's unit, that responsibility fell on him and Chase. The irony is these messages never came at a good time. It seemed they always came in the middle of Chaplain Miller's sleep cycle, whenever that was. But by and large, he was always sleeping when he was notified. The message center would send a runner to either Chase or Chaplain Miller. If it went to Chase, he would be the one to wake Chaplain Miller. There were approximately eight soldiers who had babies born while deployed, including Chaplain Miller. With the exception of one soldier, none of them were able to go home when this occurred.

Upon receipt of the birth notification, Chaplain Miller would head to the soldier's location. He delivered this news in a tent, at the gym, and in a gun tower. On the first couple of these he told the soldier, "You need to call home . . ." and would never get to the "because they have good news for you" part of the statement. That is understandable, because getting a message to call home always brings negative thoughts to the troops. However, when he did finally get to the "because they have good news for you" part, it became all smiles. After those first two, Chaplain Miller changed his opening to, "I have good news from home. You need to call your family/wife as quickly as possible." That way, it started with smiles and not panic.

There were also the bad kind of messages: a death in the family. In these instances, the only way the soldier could return home was if the death involved an immediate family member. Grandparents don't count unless there were extenuating circumstances, e.g., the grandparents raised the soldier. In fact, his first Red Cross death notification involved a grandparent. Chaplain Miller went to the soldier's location and had to inform him his grandfather had died.

It went something like this. "The reason I came to see you is I got a Red Cross notification about your grandfather."

The soldier started losing the color in his face and said, "Did he pass?"

"Yes, he died unexpectedly of a heart attack."

The soldier started rocking back and forth a little bit and started looking around. He was looking to see if it was OK to cry a little bit. He put his head in his hands and the sobbing started. Chaplain Miller sat down beside him, put his hand on his back, and said, "I'm going sit here for a little bit and when you are ready, I'm going to pray with you." They sat there for ten or fifteen minutes and not a word was exchanged. Sometimes the best things are those things that go unsaid. It was not a time for words of wisdom or gallant talks. This was a time to just be there and be available. This young man was in his early 20s, and Chaplain Miller had watched his grandfather hug his grandson before he got on the bus to leave. There were no words at that point that would truly help. But being there and letting that person know you are there for them was priceless. After about fifteen minutes, the young soldier looked at him and asked, "Can we pray now?" It was hard to determine if he really wanted to pray or if this was his way of saying, "Thank you for being here."

After praying with the soldier, Chaplain Miller said, "Come with me. We are going to my office so you can call home. There will be no time limits on this call." This allowed the soldier an opportunity to have some solitude as he talked to his family and to be able to freely grieve with them. He talked for over an hour with his grandmother, his wife, and his parents. When he walked out of the office, the color had come back to his face and his eyes were clear. For that hour, Chaplain Miller sat outside the door and prayed for him.

These notifications were so different from past wars. During WWII, Korea, and Vietnam, there was no picking up the cell phone and calling home. Red Cross notifications might take weeks before the soldier was notified of any sort of good news or misfortune at home. Rarely did a soldier get enough notification to even think about getting back for a funeral, much less the birth of a baby. Soldiers had to deal with those situations in the field and press on with what was happening around them. Technology of today has allowed instant communication around the world, and these notifications can be delivered within hours of any sort of situation occurring on the home front. If a loved one dies, the soldier can literally be home within twenty-four hours of the death. This makes the notification role of the chaplain that much more important.

Then there were the notifications that didn't come through the Red Cross or official channels. These were the notifications that came from first sergeants or commanders and mostly involved communications commonly referred to as "Dear John" letters, letters involving a wife, husband, girlfriend, or boyfriend. Chase would show up on his doorstep with this kind of news.

"Just heard from PVT Snuffy (fictitious name). His wife sent him an email telling him she was leaving, cleaning out the bank account, and moving in with Joe Blow from down the street." Unfortunately for Chaplain Miller, there were so many of these he lost count.

Before leaving, every soldier gave their spouse or significant other power of attorney over their everyday lives. This was required in order to keep households flowing and functioning smoothly. This power of attorney also enabled those significant others to do nefarious things to the soldier without their knowledge. With that understanding, the first thing Chaplain Miller would advise the soldier to do would be to cancel the power of attorney and get in touch with a representative of the Judge Advocate General (JAG) to start legal action to protect themselves. He had to explain to them that while the soldier was there, the significant other (90% of them were wives or girlfriends) was having a good time at the soldier's expense. He liked to explain the term "funding" to them. "Your (whatever) is home right now having fun, and you will get the ding later." He wanted to help where he could, as with the soldiers he counseled earlier. However, there were many cases where the spouse had made their decision and so Chaplain Miller was focused on taking care of his troops.

These notifications were different in another way, as well. With a bad Red Cross notification, there was grief and sadness. With a "Dear John" letter, the soldier was angry. They wanted to hurt somebody and hurt somebody now. Chaplain Miller used a technique to help them with this aggression. He used a workout with the soldier that would expend all of the energy and anger and advise the soldier to do this workout as often as they could until most of the anger had subsided, and it worked. This did a couple of things besides letting the soldier work it out in the gym. It also helped so that when they got into a gun truck, tower, or convoy, they would be focused on what they were supposed to be doing instead of looking to blow everything off of the face of the earth. Also keep in mind these guys walked around 24/7 with lethal weapons at the ready. By working out the aggression through a physical workout, they didn't get upset with the folks they were working with. Nobody wanted a gunfight because someone received a "Dear John" letter.

There was one funny incident that happened after one of these workout sessions. After they had finished punishing their bodies, Chaplain Miller asked if he could pray with the soldier. He said, "No, I would like you to pray for my future ex-wife." "Sure. How would you like me to pray?" Chaplain Miller said. "Can you pray she gets hit by a truck?" "NO! I can't do that!!" Jeremy laughed. "Then can you pray that something happens that makes her as miserable as me?!"

This was the gamut of issues chaplains deal with on an everyday basis. It shows chaplains cannot come into a situation with a set playbook. They have to be versatile and flexible in how they handle certain situations. One method of dealing with a soldier and a particular situation might not work with another soldier facing the same thing. Chaplains have to be able to read their troops and be able to apply a variety of different solutions. And sad to say, sometimes there simply isn't anything a chaplain can do to help a soldier. Sometimes nothing works and those situations can get extremely messy. It didn't happen on this deployment, but there have been instances of one soldier shooting another because of a bad situation at home. Then the chaplain has to deal with that, as well. It is not an easy job, but then again, God never promised us it would be easy. ***"I have said these things to you, that in me you may have peace. In the world you will have tribulation. But take heart; I have overcome the world."*** John 16:33 (ESV)

Chapter 43

Sights and Sounds

The sights were different in the air than from the ground. You may say, "Duh!!" but maybe I can explain why it was significant. From the air, everything was peaceful, looked great, and was beautiful. As they flew over the desert, they saw amazing sunsets and sunrises. It was hard, when looking down from five thousand feet, to keep in mind there was a war going on. It looked like flying over any other country.

On the ground at Bagram, there were no illusions about war. It was evident everywhere they looked. It was a military airfield and operating base for all of the different branches of the military operating in Afghanistan. Looking into the sky, all sorts of military aircraft were visible, from drones and helicopters to fighters and transport aircraft. On the south side of the airfield, there was a huge compound composed of extremely large tents housing units as they processed both into and out of Afghanistan. Bagram had a road right down the middle, called Route Disney. It was named after Specialist James Disney who died at Bagram in 2002. This was the main traffic flow, and everything worked off of Route Disney. The only time Disney shut down was for a unit to do PT in the morning. Other than that it was always busy. To the northwest sat the mountains, huge snowcapped mountains. The wind blew constantly, it was just a matter from which direction.

Near the main entrance to Bagram was a massive bazaar, a hub of Afghan tents and buildings selling everything under the sun. There were a number of dining facilities offering a variety of foods. One of Chaplain Miller's favorites was a BBQ dining facility that served a tolerable type of BBQ. Being from the South, of course, it was not like home but, again, was tolerable. The worst part was not being able to get sweet tea with it.

All of these facilities were set up in large white circus tents. Now, they weren't literally circus tents but were that large and had that look. Chaplain Miller even commented on how he felt right at home inside one based on his upbringing. One of the strangest sites was the usage of the CONEX containers. They were used for everything: storage, living quarters, operations centers, and shops inside the bazaar.

Kandahar, two to three hours by air to the south, was very different from Bagram. Where Bagram was kind of rocky due to the mountains, Kandahar was sand. Sand was everywhere and got into everything. The Americans were not in charge of Kandahar; it was controlled by a European conglomeration. Signage had a European flair to it. There were a lot of different nationalities represented through contractors and different agencies. Each nation had its own separate compound, and within each compound was a dining facility featuring their country's cuisine. Chaplain Miller always made the effort to eat at one of those dining facilities whenever he was in Kandahar.

Kandahar was more spread out than Bagram. At Bagram, it had a large city feel with people kind of stacked on top of one another. At Kandahar, it had more of a larger city feel. It had all of the communities with all of the different ethnic representation. Imagine New York City with its various ethnic communities, and this was representative of Kandahar. In addition to the European countries, there were nations from other parts of the world based there, e.g., Jordan and Armenia.

There was another location named Bastion that was similar to Kandahar except instead of buildings, everything was tents. At that time the U.S. Marines were in charge, and true to Marine fashion, it was a more organized location. This is also the location where military working dog training was conducted. There was a military police unit attached to the Arkansas deployment stationed there, which gave Chaplain Miller a reason to visit this location. He had a lot of fun visiting here.

Being a military working dog facility, there was always some kind of training happening for the dogs. Whenever Chaplain Miller would arrive for a visit, they let him get involved in the training. He and Chase would dress in bite suits and let the dogs attack them. A bite suit is a suit an individual wears that allows the dog to attack them without biting all the way to the skin. It is a thick, fiber filled suit that wraps around the torso, arms, and legs. The head is protected with a football-like helmet. Once suited up, they would hide, and the handlers would send the dogs on a search mission. When the dogs found them, the dogs would attack until given the command to cease. Fortunately, none of them ever got through the suit.

Jalalabad was an anomaly to the other bases. It was a major city with a major thoroughfare dividing the base from the city. What that meant was there was a contingent of insurgents living within hundreds of feet from the base. The base was a staging area for pushing out into the city and the countryside. The primary fortification for the base was a device called a Hesco. A Hesco is a large wire box with heavy duty material fashioned around the inside that is placed side-by-side with other Hesco boxes and filled with dirt, rocks, or other material. It is designed to stop all types of rounds. They are approximately nine feet tall, ten feet long, three feet wide, and can be stacked one on top of another providing a significant barrier against attack. Build guard towers at intervals and you have a pretty good defensive position. Where one would hear jets and large aircraft at Bagram and Kandahar, helicopters were the primary means of air travel here. The airstrip was large enough to handle a C-130, but those were few and far between.

There is no doubt that the large FOBs like Bagram and Kandahar were huge, and Chaplain Miller only liked being at them when he had to be or if he was waiting for flights. He personally loved the smaller outstations that dotted the country. The smaller bases may have only housed a few platoons or companies. Less to see, less to do, but the troops at these locations loved the visits.

Most travel was uneventful for Chaplain Miller and Chase. One things soldiers know all too well is the old adage, "hurry up and wait." One day while traveling near some of the outposts along Afghanistan's border, the guys had to wait until nightfall before their flight would arrive. The two sat down for a meal from their MREs and realized that Chaplain Miller was without a spoon. Chase finished his food first, licked and wiped the spoon clean and handed it to Chaplain Miller. This is a normal activity for soldiers. If a fellow soldier needed something, one made every effort to help. Eating from the same utensils is not an uncommon occurrence and seemed natural in that environment.

They sat for what seemed like hours beside a grassy field that had a rough runway cut into it for supplies to be able to come in and out by air. As night set in, the plane arrived, as did a convoy that would be taking supplies out to the troops and loading material or people on the aircraft. Little did Chaplain Miller and Chase know that in this particular case, the people being brought to the aircraft were enemy combatants. These men had been recently captured, and military police (MP) were transferring them to the local Afghan prison.

Chaplain Miller and Chase boarded the plan and found a spot on the plane's floor. They weren't there long, because the prisoners were being loaded and needed that space. The prisoners were young men, some of them looking to be mere teenagers. Chaplain Miller knew that these young men were not innocent and were in the wrong place at the wrong time. They were captured Taliban fighters. The MP came to Chaplain Miller and said, "Chaplain, I hate to do this but are you OK if I put this man beside you for the flight? He is restrained and we are watching them all." They guided the man to be seated and made sure he was secure for the flight. He had eye covering and headphones to protect his ears during the flight. He could not see or hear Chaplain Miller. When Chaplain Miller looked over, he realized it was a young teenager who was scared to death. It was quite possible he had been paid to fight, as many of them are. Some sought glory while others sought martyrdom. The reasons why didn't matter; he was a captured enemy combatant.

When the plane rumbled down the runway and lifted into the air, Chaplain Miller could feel this teenager shaking more and more. He knew this was very likely his first flight, whether the young man understood he was on a plane or not. The teen began to cry a little and then got sick from the turbulence. Chaplain Miller noticed that the smell of urine increased and figured this young man was truly terrified.

Prior to this moment Chaplain Miller could not imagine what praying for your enemies truly looked like. Yet, here he sat on the floor of a plane with this boy. Rather than recoil from the smells and sight of this scared kid, Chaplain Miller leaned in so the two were touching shoulder to shoulder and began to pray. He asked that God give this boy peace. He did not know his story, nor would he ever. He knew the boy had been captured and had been one of the Taliban fighters actively seeking to kill troops. Chaplain Miller prayed for the boy to come to understand that he was safe at this moment. The troops around him were not trying to harm him and had actually done everything to protect him during this time. Chaplain Miller prayed over this boy until he saw his breathing calm and his shaking slow. He understood how we are not called to dehumanize our enemies. He knew that we must also pray for them, and this was never made clearer than on this late night flight.

By praying for his enemies, Chaplain Miller was simply following the instructions of Jesus. ***"You have heard it said 'Love your neighbor and hate your enemy' but I tell you love your enemies and pray for those who persecute you."*** Matthew 5:43-44 (NIV). Most of us don't encounter enemies who want to kill us, but this verse applies in this situation as well as any situation in which we believe we have someone who bids us ill will. In a situation where Chaplain Miller could have done anything from ignoring this individual to being extremely hateful, he chose to follow the example of Jesus. I wonder what the world would actually look like if more Christians chose to follow this example.

Chapter 44

April

April arrived and everybody was starting to see the light at the end of the tunnel. Everybody knew, or was beginning to realize, in sixty days or less they would be headed back to the states, which meant back to their loved ones. It also meant the weather was beginning to warm up and fighting season was commencing. There was more potential for encounters with the Taliban. There were more IEDs being placed and encountered, with contact and combat from those encounters increasing. There were more rockets flying across the sky, as well. It was a double-edged sword for the troops. They were anxious about going home, and their thoughts were beginning to move that way, yet they had to maintain their edge and sharpness because the mission wasn't going anywhere.

In many areas, the enemy was now refreshed with new supplies and reinforcements. In addition, all of the plans that were made during the winter were starting to be implemented; for example, increasing force protection and construction projects. With the warm and dry weather, those projects gained traction. Commanders were striving to leave the place better than they found it. It was like a mad dash to get everything done so the units could turn their full attention to going home because it's not a simple matter of getting on a plane and leaving. Just as there are months of preparation prior to deploying, there are also periods of preparation for deploying back to the States.

As these projects kicked off, the command structure would travel to observe the progress being made, which meant Chaplain Miller and Chase would travel with them and continue to walk among the troops. These trips gave Chaplain Miller one of the few pleasures he had while in-country. He loved to fly on the helicopters. He didn't mind flying on the C-130 or the other fixed wing aircraft, but flying in a Blackhawk or a Chinook was a totally different experience. If he was flying on the Chinook, he would always take the seat right next to the ramp that allowed him to see out of the aircraft and observe the countryside below. It gave him a different viewpoint of Afghanistan, and he loved to observe the beauty of the country.

Upon arrival at the outposts, they would help gather as much as they could from the troops to carry back to Bagram to prepare to be sent to the States. They would pack large, black boxes with as much gear as they would hold and put them on the aircraft. Once it started getting warmer, they didn't need their cold weather gear. By packing it up and sending it ahead of time, the troops didn't have to worry about it when it was time to leave. It made the troops lives much simpler. Upon return to Bagram, the boxes went to the logistics area to prepare for shipment.

One day while Chaplain Miller and Chase were picking up new items to take to one of the bases, one of the neatest items of the deployment arrived. It was a non-lethal device designed to keep people at a distance. If someone got too close to a tower or a checkpoint and did not stop, this device could be deployed. It was pointed at the person and emitted a high decibel sound wave that would stop a person and bring them to their knees.

Chase got a bright idea: "Let's check this out," and Chaplain Miller bought in immediately. There was supposed to be some intelligence between the two of them, but at that moment, their inner little boys came to the forefront. The old saying, "The only difference between men and boys is the price of their toys" kicked in. This was a new toy, and they couldn't wait to play with it. They asked some MPs if they would like to be in charge of the test run and being, boys at heart, they said yes. As with any nonlethal items, these MPs knew eventually they would test it far more thoroughly in order to use it safely and effectively. The group set up in a small field on the base away from everybody.

Once set up, they began reading the instructions. One of the MPs read that if an individual is within fifty feet of the device, it shouldn't be turned up above a certain decibel level. On the first test, Chase was the guinea pig, which was only natural since it was his idea in the first place. He walked out to the far end of the field and began walking back toward the group and the device. They put the device on a very low setting, and at about twenty-five meters, Chase let them know he was feeling the sound wave. The MPs were satisfied the device worked properly, but Chaplain Miller smiled and asked if they needed to run any additional tests. Chase, feeling confident from the previous round, ran back to the end of the field and told the MP to turn the device up about halfway. Chaplain Miller yelled at Chase, "We're going turn it up," and Chase yelled back, "I can take it, let's do this!" and began walking toward them. As Chase was walking up, Chaplain Miller whispered to the MP, "When he gets about twenty-five meters out, crank it up a couple of notches," which he did. Chase dropped to his knees and they immediately shut the device down. When asked what it felt like, Chase responded, "The vibration in my body was tremendous," and then said with a grin, "Let's do it again and turn it full on. I want to see what it will do." Only the mind of a GI. Luckily, cooler heads prevailed at this point, or so Chaplain Miller thought. He went to get some coffee for everybody, and as he was walking back to the area, he heard one of the MPs make a really bad noise, and he hurried back to the group. The soldier was grinning and said, "I got to nine!"

The next day, they took the device to one of the outposts on the far western part of Afghanistan. Upon arrival, they discovered a unit training some new K-9 units for deployment. Chaplain Miller asked how they could help so they dressed him and Chase up in bite suits and put them in hiding places. The troops were pleasantly surprised, to put it mildly, that the Chaplain would put on one of these suits and let dogs attack him. This unit was using Belgian Malinois, which are a little smaller than a German Shephard but can be stronger. They are also very intelligent dogs, which makes them very good for this job.

The area they were working had all kinds of equipment and vehicles. They allowed the dogs to catch their scent and took off in different directions. Chaplain Miller hid in a Humvee. Soldiers were lining the area and having a ball watching the dogs track their chaplain. Here were the guys who normally bring them coffee and deal with their issues hiding in vehicles waiting on dogs to attack them. Suddenly, Chaplain Miller heard the same noise from Chase that he made when they were testing the sound gun on him. The dogs had found him and were dragging him out of a vehicle in a not so kind manner. Out of nowhere, a dog was coming in through Chaplain Miller's window and clamping down on his arm, trying to get him out of the vehicle. Chaplain Miller pulled his arm back and the dog released the grip. He thought, "Ok, that wasn't bad," until the dog clamped down again with a deeper bite. Then different thoughts went through his brain. Chaplain Miller managed to get the door open and they tumbled out of the Humvee. The dog began dragging him across the ground and there was nothing this 6'1", 215-pound well fit soldier could do about it. The troops were howling with laughter and he was laughing as well because he was having a blast. The trainers called the dogs off and gave them treats for doing such a good job. Then Chaplain Miller said the most unexpected thing. "Let's do it again." The trainers said most folks don't go twice, but he and Chase were ready to go again and did so.

This time the scenario involved trying to run from the dogs. The trainers told them to take off running and the dogs would hit them either on the shoulder or the back of the leg. They took off running as fast as they could in those suits, and suddenly, Chaplain Miller felt four paws hit him in the back and watched as the dog clamped down on his shoulder. The animal hit him with such force that he had no choice but to fall face down onto the ground, and once again, here they went with the dog dragging him across the ground.

After getting the suits off and regaining some composure, Chaplain Miller asked the guys if he could share some Scripture with them. The response was amazing. Not a single soldier left, not even the ones who never came to his Bible studies on other days. Normally, he would only get three to five soldiers at those times. Today, he had between thirty and forty soldiers listening to him, all because he was willing to put himself out there in their environment and showed he was interested in them. It was a perfect example of how to connect with individuals. By going into their environment and interacting with them in ways not normally seen, they were willing to trust and listen to what he had to say.

By showing an interest in the individual and what that individual does, one will have a better chance of sharing the good news of Jesus. Jesus didn't win people by preaching in churches. He changed lives by going to them and showing people He was interested in them. It is the main premise of the Great Commission. Chaplain Miller was not reaching these troops by sitting in an office or speaking from behind a lectern, he was reaching them by going to them, being with them, and showing an interest in them.

After he finished speaking with the troops gathered there, he started to walk away. A couple of the soldiers approached him and asked if he wanted to see where the dogs were kept and kenneled. They began to walk towards the kennels and it became apparent these guys did not want to talk about the dogs. He was able to counsel in private with these guys about things in their lives all because he was willing to let dogs "safely" maul him and drag him across the ground.

It is amazing how a little thing will open up the opportunity to speak and witness to someone. This wasn't the only thing they did with the troops. Chaplain Miller and Chase would help put up new lights, work with a crew putting in new concrete, and tasks like that to make things easier on the troops. It also opened up many opportunities to speak with the troops in their environment. Some of this was as a result of the rocket incident at the chapel and Jeremy's realization he was not supposed to spend his days sitting at the base, but needed to be out among the troops.

There was one more memorable moment during the month of April. They had been to an outpost outside of Kabul and were returning to Bagram in a flight of three Blackhawks. It was in the middle of the night, and Chaplain Miller was settled in and enjoying the peace at that time. Suddenly, the Blackhawk he was riding in began to take evasive maneuvers. The aircraft banked hard to the right, back to the left, and dipped and climbed. It was apparent the flight was under attack and the lead aircraft was the target. The gunners in his aircraft dropped their night vision goggles (NVG) into place and prepared their weapons to fire. They were searching the ground below for incoming rounds in order to return fire. That feeling of "what can I do" came over Chaplain Miller, and just as quickly, the answer of "you have zero control here" came over him. They started circling and circled for about an hour. Chaplain Miller, with his NVGs, could see a convoy underneath them and that convoy was in contact with the Taliban. The Blackhawks stayed on station until the convoy issued the all clear and they resumed their flight back to Bagram.

This wasn't the first time Chaplain Miller had been on an aircraft that had taken evasive action. The first time, they were on a C-130 flying back to Bagram from Kandahar. The aircraft had just cleared the mountains and was descending in altitude. From a hidden position somewhere in the rocks, a rocket was fired at the plane. As they were on their landing approach, the aircraft began to take evasive maneuvers and began firing flares in order to fool what might have been heat seeking missiles. He was sitting beside Chase in the jump seats. The crew chief jumped up and started moving about the aircraft and was listening to what the pilots were saying over his headset. Suddenly, he yelled, "Hold on to what you got because we are going to be moving around a lot." When he finished, Chase reached over and put his hand in the middle of Chaplain Miller's chest like a mom would do with a child in their car. "What are you doing?" Chaplain Miller yelled. "I don't know! I'm protecting you?" Chase yelled back. "So what is your hand going to do if this aircraft goes down?" They both started laughing as the plane moved violently across the sky. Either the rocket ran out of fuel or couldn't keep up with the maneuvers, because the plane eventually landed without incident. When they got into the vehicle to take them back to their quarters, Chaplain Miller asked him, "So just what did you think you were going to do back there?" Chase replied, "I don't know, but I had to do something. It is my job to protect you."

It was an incident that made Chaplain Miller realize that he really wasn't in control of very much. He definitely wasn't in control of that aircraft and he wasn't in control of the Taliban as they tried to shoot at the aircraft. It's one of the moments when one realizes they really aren't in control. It's the one thing God puts on your mind to help you realize that He is the only one in control, and that is where your trust should go. ***"Trust in the Lord with all your heart and lean not on your own understanding; in all your ways acknowledge Him, and He will make your paths straight." Proverbs 3:5-6*** (NIV)

Chapter 45
Preparation

Moving into April, fighting season hit full force. Outposts all around the country began receiving rocket, mortar, and small arms fire, as well as the occasional convoy attack. The Taliban were coming back through Pakistan, their forces reinforced over the winter, and they were active. On our side, patrols became more active, and QRF were on higher alert and more supportive of these patrols and convoys. All in all, it was a major uptick in operations and conducting business. Add to this, there were also less than sixty days until it was time to go home, and that was never far away from anyone's mind. This was also the time when the command structure began to hammer home situational awareness and readiness. They still had everybody and wanted to make sure in these last couple of months it stayed that way. This was not the time to let their guard down.

It was with this thought in mind that the commander notified Chaplain Miller that when the time came, he would be on the first flight out of Afghanistan. It would go to Manas, Kyrgyzstan, an intermediate stop before heading further west. The commander told him, "I'm sending you first for a specific reason. I want you to put yourself in front of every member of our deployment as they come off of their planes. You'll have enough time between flights to visit with each soldier, and I mean every one of them. Check on how they are doing, find out what their plans are when they get home, and try to discuss any issues they are having or expecting. In addition, I want you to have a class for each flight of soldiers after they arrive." Chaplain Miller confirmed this by saying, "OK, let me make sure I have this right. You want me to meet individually with each of the three hundred soldiers deploying home and also to have a class for each flight of soldiers after they arrive at Manas." The commander replied, "You got it. Pretty much when we hit the ground there, you aren't going to get much, if any, sleep. Get us to the finish line." The commander knew that the transition home can be tough on troops, and he was sending his best tools to deal with the issues.

With these orders in mind, Chaplain Miller and Chase amplified their trips to other stations. Every day, they were someplace new, visiting, counseling, and continuing what they started ten months ago when they first entered the country. They were sprinting towards the finish line. This sprint didn't mean other things went by the wayside. They were still coming back to Bagram to conduct Sunday services, but once those were completed, away they went again. What they started to see was the fruits of their labor. Soldiers who had been working through their spirituality began to fully embrace it. Teachings and Bible studies became even better attended. They began to realize what Chaplain Miller had been preaching was real. It was not only real in Afghanistan at the FOBs or outposts, it was real everywhere they went. This would travel home with them and be there as they transitioned back to civilian life. ***"For if you possess these qualities in increasing measure, they will keep you from being ineffective and unproductive in your knowledge of our Lord Jesus Christ."*** 2 Peter 1:8 (NIV)

Chaplain Miller introduced a new theme in his discussions, talks, and messages he had found in the writings of Paul Tripp, a religious writer. Miller discussed the difference between temporary and permanent relationships. It would go something like this. "There are differences between our relationships on this earth, but they fall mainly into two categories: temporary and permanent. For example, a relationship between a child and parents is a temporary relationship. The child is going to always love and respect their parents, but the relationship is only temporary in terms of authority. A spouse is a permanent responsibility. If done according to the Scriptures, we have a permanent responsibility to our spouses that will only end when one leaves this earth. There are all sorts of earthly temporary human relationships but the one permanent constant relationship, other than with a spouse, is the one you will have with God. This comes about through our relationship with Jesus Christ. So with that established, as we wind down, your activities need to show the importance of these permanent relationships. Your spouses haven't seen you in a year, and now is the time to make sure you are making your phone calls home and stoking that relationship instead of spending time doing other things."

This type of conversation with the troops was to help them prepare to go home. Chaplain Miller was stressing to them in these last two months. Not only should they be preparing to leave Afghanistan but also preparing to walk back into the lives they left a year ago.

One of the primary differences between the National Guard and active duty troops is money. When an active duty soldier leaves Afghanistan and returns to their base, nothing in their finances changes. When a Guard soldier returns home, their active duty pay and allowances come to a halt and they return to their civilian lives. For a large percentage of the soldiers, they are making more money on active duty than in their civilian jobs, particularly in rural states like Arkansas. In order to help them prepare for the transition back to the States, Chaplain Miller also counseled them about money, preparing to transition home, and this paycheck coming to a halt. He did this same type of counseling when they first deployed, but now is when most of the soldiers actually began paying attention.

Using these counseling sessions began to show fruits in the spiritual realm. As well as they had been relating to Chaplain Miller, because of the counseling concerning life (relationships, money, family matters, etc.), they began relating to him in a more spiritual sense. Bible study attendance grew across the board and the number of personal counselings amplified. The speed of the ministry (circulation among the troops) increased. They had seen and experienced their mortality and things were suddenly much clearer about what they needed to do. It was a perfect combination that opened the door for Chaplain Miller to ramp up the work God was putting before him.

This was also the time Chaplain Miller began communicating with the chaplain who would be replacing him. He was determined to ensure the incoming chaplain would not have the same experience he had upon arrival. He and Chase wrote new Standard Operating Procedures (SOPs) and other documentation and communicated with the incoming chaplain daily by email to make the upcoming transition as smooth as possible.

Also during this time, the advance party for the unit replacing the Arkansas unit arrived in-country and it was a mild surprise. They were all female. Most of the command positions in the new unit were occupied by female soldiers. I know in this day and time this should not be a surprise, but in a male dominated arena, like the military, this is a rarity. They had the red carpet rolled out for them, something that didn't happen when Miller's Arkansas unit arrived. They were taken to all of the outposts and FOBs and introduced to their counterparts in the field. It was as if the entire unit was getting rid of a bad experience with one sweeping move. There was also a bit of change to the mission of the incoming unit. Where the mission of the Arkansas unit was to increase security and build some places up, the incoming unit was being tasked with shutting some places down.

One thing Chaplain Miller and Chase did at the end of April was to begin to completely empty the CONEX holding the coffee and other items in it. It was about three fourths empty by this time. They used the last month and few days making sure the troops had everything they needed or wanted, and when the deployment was over, the CONEX was empty. Yes, it would have been a huge blessing to hand off to the next chaplain, but their team had worked hard, and they knew an empty container reflected their circulation and ability to walk among the troops.

Chapter 46

May

The last month was spent making sure as many of the troops' needs and issues as possible were met before they started the movement home. The reason was once they got home, Chaplain Miller would no longer have a captive audience and it would be much more difficult to get to the troops once they scattered. He would ask questions of the troops upon arrival at each location and he and Chase would make note of the ones whose answers concerned them. Upon arrival at Camp Shelby, they would spend additional time with that soldier.

The commander came to Chaplain Miller the first week of May and brought all of his travel to a halt. He informed Chaplain Miller it was time to prepare to leave for Manas. Miller and Chase began to pack up what they had and prepare for departure. Elements of the replacement unit began arriving about a week later, and the commander moved the entire Arkansas contingent from their plywood structures back to the tents. This caused a small bit of consternation about safety, but as had been proven during the rocket attack, those "hardened structures" weren't so hardened. It was like comparing a thin book to a piece of paper. Not a lot of difference in protection.

Chaplain Miller, Chase, and the other troops living in their quarters packed up. Chaplain Miller packed three bags, and everything left over was mailed home. This was pretty much the case for everybody. Then they began to "clean" their quarters. As Chaplain Miller put it, "We were dudes and had been living like dudes for the past twelve months." Make of that what you will, but needless to say, their quarters needed cleaning. The CSM emphasized to them these quarters were about to be occupied by the next unit and he didn't "want them thinking they were replacing a bunch of pigs." Their cleaning primarily consisted of a small amount of sweeping, wiping, and taking the linens that were on the beds, trading them in, and putting new linens on the beds. Dude method of cleaning.

The replacement leadership arrived and there were hugs all around, even between Chaplain Miller and his replacement. His replacement asked, "Why are you hugging me?" to which Chaplain Miller replied, "You'll understand later." They walked the new commander and her staff to their quarters and proudly opened the door for them. The new commander went into the quarters and came out. She looked at one of her sergeants and said, "We are going to need Lysol, cleaning supplies, and some brooms." It was like a kid thinking they had cleaned their room and then their mom arrived.

Chaplain Miller, upon seeing his replacement, felt his entire body breathe a sigh of relief. Even though there were still days left in-country, in his mind things were over. There was a small release of the pressure he had been working under for the past few months and he could really see the end was in sight. He took his replacement to the chapel, showed him around the facility, and basically gave him a logistical view of the land. He then took him into his office and got him acquainted to the setup. He showed him the passwords for the computer and also the phone that allowed him to call home. In fact, Chaplain Miller asked him if he had called home yet and when his replacement negatively replied, Miller told him to make that call now, and stepped outside. When the door opened, Chaplain Miller could tell he had been wiping his eyes. He reassured him that he had done the same thing and it was nothing to be ashamed of. "We are blessed to have the families we do, and missing them is normal." Jeremy said.

Later that day, Chaplain Miller and his replacement were sitting in the annex and Chaplain Miller said, "Just ask me the questions you want to ask." The replacement chaplain opened up a small notebook and began to ask a ton of questions. Chaplain Miller answered them all. When finished, Chaplain Miller said to him, "The most important thing is what your chaplain assistant is learning right now and that is how to let you travel. Do not try to arrange your own travel. Let your assistant do it for you. Lean heavily on him. Even though he hasn't been through any assistant courses, it doesn't matter. Chase didn't go through a single course and doesn't know anything about being a 56Mike (Chaplain Assistant MOS). But he's been killing it for the past year and I would put him over every CA around. He has been my protection through the entire deployment. I know you are scared, but try to look at it like I have looked at it. This is a mission field that God has given to you. You are a missionary on a one-year mission trip. Your troops are priority number one and you take care of them first. Everything else is second to that. And don't write your own awards packet." At that point, the replacement got a weird look on his face. Chaplain Miller continued, "I actually encountered that upon my arrival. If you deserve it, they will make sure you get it. Don't chase it. It is not that important. Pray you never have to go to the hospital and see soldiers who have been wounded and are missing limbs, but be prepared to. Pray you never have to do a ramp ceremony, but be prepared to. I can't give you any better advice than this."

A few days after moving into the tents, Chaplain Miller was summoned to attend a unit formation. This was one of several formations the unit conducted prior to leaving. The leadership of the unit wanted to present all of the end of tour awards earned while in-country. Chaplain Miller wasn't sure why he was being summoned to this particular formation until the commander called him forward and issued him a Combat Action Badge (CAB) earned during 20 March rocket attack. Chaplain Miller thought as the commander was pinning this CAB on him, "This is the same man who had multiple vehicles blown out from underneath him in Iraq and yet was taking care to make sure each of his soldiers was recognized." He looked at Chaplain Miller and said, "You know you are probably the only chaplain in Arkansas who is going to have one of these." It brought Chaplain Miller back to the guy he replaced chasing all around Afghanistan trying to get one, and Chaplain Miller earned his simply because he was ministering to the troops. He also felt this legitimized him not only as a chaplain but as a soldier, and that was important to him. He loved being a soldier, and he felt this gave him footing in that department.

For the next two weeks, he drove out to the circus tents where most of his guys were billeted to visit with them and make sure they had everything they needed. He would make sure they could get to the gym or to the exchanges. He continued to do the ministry of simply walking among the troops and trying to make every connection he could.

At the end of that two weeks, Chaplain Miller received word that he and SFC Jeu would be on the first flight out of country, which is exactly what the commander wanted. The next day, they began that process, and it took them about twenty-four hours to get out-processed and actually leave. As he stated, "They get you in quickly but take their time sending you home."

Chaplain Miller would be functioning as the Officer in Charge (OIC) upon arrival at Manas. Chase was deliberately placed on a later flight in case any sort of chaplain duties were required. Chaplain Miller was not comfortable being one of the first ones out and would have been preferred being one of the last. But as all military individuals know, your preferences don't matter. The commander explained to him that he wanted every soldier pushed in Chaplain Miller's direction upon arrival. He stated, "It is more important to me that you have face time with every single soldier before they leave for Camp Shelby and real life hits them square in the face. That will be when families arrive, and we will begin to lose them at that point."

The evening arrived, and Chaplain Miller and SFC Jeu went to the out-processing center and started the process of leaving the country. They were only allowed two bags to carry home (everything else was shipped). It was kind of like coming into a country with strict entrance policies. There were stations to ensure you had all of the proper paperwork and credentials. There was a bag check area, and it was stringent because it had to be. GIs are crafty and will try to bring all sorts of things home they know they can't. They have been known to take AK-47s apart and store the different parts in different bags. They try to sneak through all sorts of contraband and things that can't be described. So, bags are run through scanners and thoroughly examined, looking for anything and everything. This made for a very slow-moving process.

Once this part of the out-processing was complete, they were moved to a containment area. This basically consisted of a chain-link fenced area out near the flight line. Upon arrival, they were told they would be there about an hour. Six hours later, they were still in that containment area. Chaplain Miller was looking around, taking in the sights for the last time. He stared at the few stars he could see and at the mountains off in the distance. As daylight began to arrive, the base was starting to awaken. He heard aircraft beginning to spin up, and Blackhawk helicopters began to fly overhead headed to destinations somewhere in the country.

Suddenly he became aware he was sitting out in the open. He looked at SFC Jeu and asked, "Just where would we go if a mortar or rocket attack happened right now?" SFC Jeu said, "Well, I guess we would huddle up like we did that time before and hope it doesn't hit us." They both laughed about that and moved on to other topics.

Finally, they were called, got their bags together, and were taken to what looked like a really bad airport terminal. They walked through the terminal and were loaded onto a C-5 cargo plane, the largest plane in the Air Force inventory. As they were walking through the terminal, on to the tarmac, and to the aircraft, Chaplain Miller found time to speak to each of his soldiers on this flight. These were preliminary conversations designed to indicate if he needed to spend more time with one of those soldiers later on. As he got onto the aircraft, he had two distinct thoughts. The first thought was, "I'm not going to miss this place at all," followed immediately by, "No, I may really miss this place."

These thoughts are not uncommon among troops leaving a deployment. They are so ready to go home they truly hate where they are. This place they hate so much also provided them with a camaraderie they will never forget or possibly ever have again. For the past year, there was no thought about wardrobe or what to wear. There was something to accomplish each and every day. There were no decisions about meals or where to eat them. They haven't paid bills for a year or even had to think about them unless they got a call from home. They don't have to prove themselves to anybody or worry about what someone thinks. It's been an entire of year of "bro time," and those times will be missed. The reality of the situation is the relationships a soldier, airman, sailor, and marine has had during the deployment change the instant they set foot out of country.

The plane departed Bagram and two hours later landed in Manas. When they got off of the aircraft, the first thing Chaplain Miller noticed was it was all green. There was grass and trees. As the bus took them to their quarters, they were informed about the places to relax and unwind. There were twenty-four hour movie theaters and restaurants with big screen TVs. There was a twenty-four hour exchange with ice cream machines and candy bars as well as other things they hadn't seen in a year. There was a massive gym.

Chaplain Miller found the chapel and was in awe of all the stuff they had to give away to the soldiers. The resident chaplain gave him a cold drink and took him to a room that was silent. There was no generator hum, no aircraft sound, no radios; no sound of any kind. He used this time to call Beth and let her know he was out of country, and then he went back to his troops.

Chaplain Miller had a primary message he wanted the troops to hear. He had three things he wanted to do with them as they arrived. He wanted them to understand what right relationships look like. They were going to hear about where to really turn to when they need help. And lastly, he wanted to pray with each of them. This was the message he conveyed to every one of his soldiers over the next three days.

It was during this time that a number of troops told him thank you for what he had done. More than one soldier told him the advice he gave about how to talk to their spouse helped save their marriage. Troops kind of let their hair down and really opened up. It felt good for Chaplain Miller to see the fruits of his labors over the past year. It took him back to Scripture in 2nd Peter where Peter says we need to be an example even to unbelievers.

He was also told that he had made an impact on believers and helped them grow closer to God. By listening to his devotionals and prayers with them and chewing on those words, it allowed the Spirit to work on them, moving them to be better with their Bible study and prayer. They weren't all positive responses but the positive ones helped buoy his spirits and reveal that God does work through us when we allow Him into our lives.

It took three days for his unit to fully arrive at Manas. During that time, he did exactly as the commander had tasked him. Every soldier met him face-to-face, including the commander and the CSM. They may have wanted to brush him off at first but quickly relented knowing he was doing his job and caring for their souls. Early the next day, Chaplain Miller was standing on the flight line looking at the plane that would take him home. After boarding and settling in, he passed out. Waking up several hours later, he recalled the words of the pilot over the radio. "Soldiers, thank you for your service. You have now entered U.S. airspace and in a few minutes will be flying over U.S. soil. Welcome home!"

Afterword

God was not through with Jeremy when he stepped off of the plane and was home. He has continued to develop him and use him to impact the lives of countless troops and those around him. Since arriving home, Jeremy and Beth have accomplished the following.

- Beth and Jeremy took on the hobby of renovating homes to rent or sell. Together, they have worked on 12 home projects.
- The couple has continued to host exchange students from Germany, Nigeria, Brazil, and Taiwan.
- Jeremy completed his Doctorate Degree from Midwestern Baptist Theological Seminary. His graduation project consisted of a Biblical guide to help churches and/or chaplains counsel someone dealing with thoughts of suicide.
- Jeremy was promoted to the rank of Lieutenant Colonel and has been selected for the rank of Colonel to be assigned as the Joint Forces Headquarters Senior Chaplain for the Arkansas Army National Guard.
- Outside of his military duties, he serves as an adjunct professor of theology for Central Baptist College in Conway, Arkansas, and as a fitness coach for teenagers at CrossFit Greenbrier in their hometown.

This is just another example of how God continues to work in us, and when one mission field is complete, He is preparing us for others. Jeremy's story, though different from most of ours, is actually the story God wants for all Christians. Matthew 16:24-26 (NIV) says, ***"Then Jesus told his disciples, 'If anyone would come after me, let him deny***

himself and take up his cross and follow me. For whoever would save his life will lose it, but whoever loses his life for my sake will find it. For what will it profit a man if he gains the whole world and forfeits his soul? Or what shall a man give in return for his soul?'" This is not just a suggestion but what every Christian should take to heart. Once we surrender our life to Christ, it is ours no longer. It belongs to Him, and we follow where He leads. Even if it means going to a war zone.

Made in the USA
Monee, IL
04 November 2026

67383841R00148